SPECIAL PAPERS IN PALAEONTOLOGY NO. 3

UPPER MAESTRICHTIAN RADIOLARIA OF CALIFORNIA

BY

HELEN P. FOREMAN

With 8 collotype plates, and 1 text-figure

THE PALAEONTOLOGICAL ASSOCIATION

LONDON

SEPTEMBER 1968

PRINTED IN GREAT BRITAIN

CONTENTS

ABSTRACT. Seventy-seven species of Radiolaria (fifty-six of them new) have been described from four localities of the Moreno formation in Fresno County, California. Samples from one of these localities, a core, extend into probable Paleocene. In addition to the samples used for descriptions of species, other Maestrichtian to lower Santonian samples from California, the Caribbean, and the South Atlantic have been examined to determine time-ranges. Of the described species, only five extend beyond a short transitional zone up into probable Paleocene. Four have been recognized in the oldest sample studied from the lower Santonian of Trinidad. Eight new genera, *Acidnomelos*, *Calyptocoryphe*, *Cornutovum*, *Ectonocorys*, *Eribotrys*, *Gongylothorax*, *Myllocercion*, and *Solenotryma*, are described. Special attention has been given to describing and illustrating details of cephalic structure as it is believed these are important in establishing relationships and understanding evolutionary changes which will increase the usefulness of this group for stratigraphic correlation.

INTRODUCTION

THE object of this paper is to report some of the most characteristic or abundant species of radiolarians in the upper Maestrichtian of California, and to establish whether they range into Paleocene on the one hand or further down in the upper Cretaceous on the other. The only post-Maestrichtian samples used in this study are from a single core in California, and the samples older than upper Maestrichtian are from California and the Caribbean. One submarine sample, judged to be Maestrichtian-Campanian on the basis of Radiolaria, is from the South Atlantic.

It is anticipated that this work will be continued to form a basis for the future application of the radiolarians to problems of Mesozoic stratigraphy. Although a considerable number of descriptions of radiolarian assemblages of various Mesozoic ages are available in the literature, these are not yet sufficient to indicate to what extent this group of microfossils can be used for age assignments, except in some parts of the Cretaceous in the U.S.S.R. (Lipman 1952, 1960, 1961, Aliev 1965, and Koslova and Gorbovetz 1966). Virtually all of the work on Cretaceous radiolarians in the U.S.S.R. has been on assemblages older than Maestrichtian, only two papers (Lipman 1952 and Koslova and Gorbovetz 1966) known to the author containing a few data bearing on the results of the present study.

Part of the difficulty of evaluating the stratigraphic applicability of Mesozoic radiolarians results from the fact that Mesozoic species have in the past been assigned uncritically to Cenozoic genera and families, simply because of superficial similarities. It is now becoming evident that most of the Cretaceous species are not closely related to Cenozoic forms, and should be widely separated from them taxonomically. This accounts for the large proportion of new genera, and tentative generic assignments, in the taxonomic section below. Generic assignments of the stichocyrtids have been particularly difficult. For the most part the solutions proposed are tentative and await a thorough revision of the whole group.

The number of species present in the well-preserved late Maestrichtian assemblages in California is too great for all to be included here, but an attempt is made to describe species representing most of the principal (and potentially stratigraphically useful) larger taxa. This has been done more effectively for the nassellarians than for the spumellarians, because of the greater taxonomic difficulties involved in the study of the latter.

Acknowledgements. This work would not have been possible without the co-operation and generosity of those geologists who provided the samples (to W. R. Riedel and to the author) on which it is based. These individuals and organizations include: P. J. Bermùdez; A. L. Brigger; W. R. Evitt; G. D. Hanna; J. D. Hays; A. R. Loeblich, Jr., Chevron Research Co.; M. B. Payne; and J. B. Saunders, Texaco, Trinidad Inc. In addition to assisting in the provision of samples, W. R. Riedel provided copies with translations of much of the Russian literature, helpfully discussed problems of generic taxonomy and suprageneric implications, and critically read the manuscript. Arthur Mack of the Oberlin College Classics Department was helpful in suggesting suitable species names. For financial support the author is grateful to the National Science Foundation, grant GB 3860.

STRATIGRAPHIC RESULTS

It has long been recognized (Squinabol 1903, Campbell and Clark 1944*b*, Lipman 1961, and Koslova and Gorbovetz 1966) that the radiolarians of the Cretaceous are markedly different from those of the Eocene, but the details of the change remain obscure because of inadequate information on assemblages of Paleocene (including Danian) age. The material available for the present study includes one series of core samples passing from upper Maestrichtian to younger beds which contain a markedly different radiolarian assemblage including one species (*Trematodiscus barbarae* Middour *in* Frizzell and Middour 1951, possibly the same as *Staurodictya quartus* Borisenko 1958) known only from the Paleocene. Drugg (1966) has found a marked change in assemblages of dinoflagellates and spores at the same level in this series of samples and considers the over-lying sequence to be Danian in age. Twenty-four of the thirty-six upper Maestrichtian Radiolaria described from this core persist into a short transitional zone (at approximately 305–15 ft.)—only one of the described Maestrichtian species (?*Stichomitra alamedaensis*) ranging upward into the probable Paleocene level to the youngest sample examined at approximately 178·5 ft. Besides this short transitional zone, indicated on the chart, there is another zone (at approximately 265–85 ft.) which, while containing predominantly ?Paleocene Radiolaria, also contains (besides ?*S. alamedaensis*) *Amphipyndax stocki*, *Dictyomitra* cf. *multicostata*, *Cornutella californica*, and ?*Staurodictya fresnoensis* along with some other undescribed Radiolaria of definite Cretaceous character. The species of this younger assemblage await further study.

The downward ranges of the upper Maestrichtian species into lower Maestrichtian, Campanian, and Santonian are also evident from the species distribution chart. In summary it might be pointed out that forty-two of the seventy-seven upper Maestrichtian species described have been found in lower Maestrichtian–? upper Campanian assemblages, and only four have been found to range down to the lower Santonian. A few of the species are known to range geographically considerably beyond the distribution of the samples studied here. *Amphipyndax stocki* (Campbell and Clark) occurs in Europe and probably the U.S.S.R., and *Gongylothorax verbeeki* (Tan Sin Hok) in southeast Asia.

SAMPLES INVESTIGATED

Six of the localities mentioned previously by Foreman (1966) are included in the list below:

1. Cima Hill core hole III of the Chevron Research Corporation, Marca shale member and Dos Palos shale member (upper Maestrichtian to Danian) of the Moreno formation; Chounet Ranch quadrangle, Fresno County, California; Cima Hill in NE. 1/4 Sec. 7, T. 15 S., R.12 E. From A. R. Loeblich, Jr. Depths given in the locality data refer to depth in hole and not stratigraphic depth, samples from 175 to 385 ft. (See Drugg 1966 for details of core hole III.)
2. Marca shale (MBP) top of Marca shale member (upper Maestrichtian) of the Moreno formation; Chounet Ranch quadrangle, Fresno County, California, Sec. 6,

T.15 S., R.12 E., 450 ft. north and 1,950 ft. west of the SE. corner of Sec. 6. From M. B. Payne.

3. CAS loc.1144, Moreno formation (upper Maestrichtian) Chounet Ranch quadrangle, Fresno County, California, Sec. 6, T.15 S., R.12 E.; loc. 3 of Long, Fuge, and Smith (1946); California Academy of Science locality 1144 of Hanna (1928). From G. D. Hanna.

4. Moreno Gulch, three composite samples are listed below. They are not distinguished in the body of the text or the locality chart. Brigger's numbers following the word 'Moreno' are used to distinguish different size fractions from the same composite locality and are included to complete the description.

Moreno 3518 and 3519, 'Y' Canyon (personal locality name of A. L. Brigger), Moreno formation (upper Maestrichtian); Panoche quadrangle, T.14 S., R.11 E. Collected, prepared by, and received from A. L. Brigger.

Moreno 3510, 3511, and 3517, Water Canyon, Moreno formation (upper Maestrichtian); Panoche quadrangle, T.14 S., R.11 E. Collected, prepared by, and received from A. L. Brigger.

Water Canyon (personal locality name of J. Smith, according to G. D. Hanna, personal communication), Moreno formation (upper Maestrichtian); Chounet Ranch quadrangle, Sec. 11, T.14 S., R.11 E.; loc. 2 of Long, Fuge, and Smith (1946). Collected by J. Smith, received from G. D. Hanna.

5. UCMP loc. A2615, 'Moreno Grande' formation of Huey (lower Maestrichtian–? upper Campanian, Goudkoff D-2); Tesla quadrangle, Alameda County, California, north central part of Sec. 31, T. 3 S., R. 4 E. Concretion in east bank of Mitchell Creek, Corral Hollow. Collected by Arthur Huey. This is the type locality of Campbell and Clark (1944b). University of California Museum of Palaeontology.

6. CAS loc. 39545, 'Moreno Grande' formation of Huey (lower Maestrichtian–? upper Campanian, Goudkoff D-2); Tesla quadrangle, Mitchell Creek, Corral Hollow, Alameda County, California. Concretions from approximate UCMP A2615 locality; collected by G. D. Hanna and C. Chesterman; prepared by A. L. Brigger. From G. D. Hanna, California Academy of Science.

7. Stanford loc. PL2875, upper part of Uhalde formation (lower Maestrichtian–? upper Campanian, Goudkoff D-2) of Panoche group; Patterson 7·5-minute quadrangle, Stanislaus County, California, just north of the centre of NW. 1/4, NE. 1/4, SE. 1/4, Sec. 32, R. 7 E., T. 5 S. From large limestone concretion exposed on south-west slope of topographic nose dividing the upper tributaries of the dry stream gully which extends southeastward into Black Gulch. From W. R. Evitt.

8. Evitt slides AE24, 26, 27, and 28, Ragged Valley shale equivalent (lower Maestrichtian–? upper Campanian, Goudkoff D-2); Patterson 7·5-minute quadrangle, Stanislaus County, California, on tributary to Black Gulch NE. 1/4, SE. 1/4, Sec. 32, R. 7 E., T. 5 S. AE24, 26, 27 from nodule, AE28 from same locality in limestone bed. These are slides of picked radiolarians and are included on the locality chart because of the information they give on the ranges of some species. However, absences of species from these slides may be due to selective picking.

9. Lamont V–18–129, 78 cm.; South Atlantic, 41° 42′ S., 56° 35′ W. (Maestrichtian–Campanian) based on Radiolaria. From J. Hays, Lamont Geological Observatory collections.

TABLE 1. Ranges of upper Maestrichtian species studied. Relative abundance in the three upper Maestrichtian localities most intensively studied is given: A—abundant, C—common, R—rare, and VR—very rare. Occurrences in other localities from which smaller samples were examined are indicated by P—present. Evitt slides are included because they extend some of the ranges; absences from them may be due to selective picking.

	? Paleocene	Transitional	upper Maestrichtian				lower Maestrichtian– ? upper Campanian				Maestrichtian– Campanian	Campanian– Santonian	lower Santonian
	Cima Hill 175–297 ft.	Cima Hill 305–15 ft.	Cima Hill 315–85 ft.	CAS loc. 1144	Marca sh. (MBP)	Moreno Gulch	UCMP loc. A2615	CAS loc. 39545	Stanford loc. PL2875	Evitt slides AE24–28	Lamont V–18–129	Cuba B191	Trinidad well Marac 1
Acidnomelos apteron				C	C	R							
Amphipyndax plousios		P		R	R	R		P		P			
Amphipyndax stocki	P	P	P	R	R	C	P	P	P		P	P	P
?Archicorys allodarpe				R		R							
?Bisphaerocephalina apimelos				R		R							
Calyptocoryphe cranaa				R	R	R							
Cenellipsis heteroforis				C	C	C	P	P	P				P
?Clathrocyclas diceros		P	P	R	C	C		P	P				
?Clathrocyclas hyronia					VR	R							
?Clathrocyclas lepta						R							
Cornutella californica	P			C	C	C	P	P	P	P	P		P
Cornutovum levigatum				C	C	C							
?Cryptocapsa asymmetros				R	R	R							
?Cryptocapsa axios				VR	VR	VR		P					
Dictyomitra andersoni		P	P	A	A	A	P	P	P	P	P		P
Dictyomitra cf. crassispina		P	P	R	C	R							
Dictyomitra lamellicostata				C	R	R	P						
Dictyomitra cf. multicostata	P	P	P	A	A	A	P	P	P	P	P		
Dictyomitra rhadina				R	R	R		P	P				
Dictyomitra regina				R	R	C	P	P	P		P		
?Dictyomitra ascelis				R	R	C	P	P					
Ectonocorys deltota				C	R	C		P					
Ectonocorys lampra				R	R	C		P					
Ectonocorys scolia				R	R	R	P	P					
?Ectonocorys hemicarena			P	C	R	R							
Eribotrys anax		P	P	R	R	R		P					
Eribotrys despoena				R	R	R	P	P					
Eribotrys litos				R	R	R							
Gongylothorax verbeeki			P	R	R	R	P		P	P		P	
Hemicryptocapsa conara			P	R	R	R							
?Lithelius robustus	P		P	C	C	C	P	P	P				
?Lithocampe eureia			P	C	C	C		P					
?Lithomelissa amazon			P	C	R	C							
?Lithomelissa heros	P		P	A	C	A	P	P					
?Lithomelissa hoplites				R	R	R		P	P	P			
?Lophophaena glaucidium				C		R							

	? Paleocene	Transitional	upper Maestrichtian			lower Maestrichtian–? upper Campanian				Maestrichtian–Campanian	Campanian–Santonian	lower Santonian
	Cima Hill 175–297 ft.	Cima Hill 305–15 ft.	Cima Hill 315–85 ft.	CAS loc. 1144	Marca sh. (MBP)	Moreno Gulch	UCMP loc. A2615	CAS loc. 39545	Stanford loc. PL2875	Evitt slides AE24–28	Lamont V–18–129	Cuba B191	Trinidad well Marac 1
? Lophophaena polycyrtis			P	C	C	C	P	P	P	P	P		
Myllocercion acineton			P	R	R	R	P	P	P				
Myllocercion rhodanon				VR	VR	VR							
Myllocercion sp.		P	P	P	P	P		P	P		P		
Pseudoaulophacus spp.				R	R	R						P	P
Rhopalosyringium colpodes			P	R		VR							
Rhopalosyringium elasson				R	VR	VR		P					
Rhopalosyringium ? magnificum		P	P	C	R	C		P		P			
Rhopalosyringium sparnon				VR	VR	VR							
? Saturnalis deiropede			P	R	R	R		P		P			
? Sciadiocapsa ptesimolecis			P	C	R	R							
? Sciadiocapsa causia				R	R	C	P	P					
? Sciadiocapsa petasus				R		R		P					
Spongosaturnalis campbelli			P	C	C	C				P			
Spongosaturnalis nematodes			P	C	C	C							
Spongosaturnalis spiniferus				R	R	C	P	P		P	P		
? Spongosaturnalis catadelos		P	P	C	C	C							
Solenotryma dacryodes				VR	VR	VR							
Solenotryma sp.			P	C	C	C		P					
cf. Solenotryma dacryodes						VR	P				P		
? Staurodictya fresnoensis	P	P	P	A	A	A					P		
Stichomitra asymbatos		P	P	C	C	C	P	P	P	P	P		
Stichomitra cathara				R	VR	R							
Stichomitra compsa		P	P	C	C	C		P					
? Stichomitra alamedaensis	P	P		VR	R	R	P	P	P				
? Stichomitra campi		P	P	C	C	C	P	P	P	P	P		
? Stichomitra cechena				R	R	R							
? Stichomitra livermorensis		P	P	R	R	R	P	P	P	P	P		
Stichopilidium teslaense		P		R	R	R	P	P	P		P		
? Tetraphormis leia			P	C	C	C							
? Tetraphormis lobocarena						VR							
Theocampe altamontensis				VR	VR	VR	P	P	P				
Theocampe argyris			P	C	R	C							
Theocampe bassilis				VR	VR	VR							
Theocampe dactylica				VR	VR	C		P	P	P			
Theocampe daseia		P	P	R	R	R		P					
Theocampe lispa				R	R	R							
Theocampe vanderhoofi		P	P	A	A	A	P	P	P	P	P		
? Theocampe stathmepora				VR	VR	C							
Theocapsomma amphora		P	P	C	C	C		P	P		P		
Theocapsomma ancus		P		R	C	R							
Theocapsomma comys		P	P	A	A	A				?			
Theocapsomma legumen		P	P	R	R	C	P	P	P	P			
Theocapsomma teren				R	R	R		P	P				
Trematodiscus barbarae	P	P											
? Tripodocorys megalocarena			P	R	R	R		P					

10. Cuba B191, 'Pre-Habana' formation of Cuba considered by Bermùdez (1963) to be Santonian–Campanian in age. Excavation 105 m. south of the church at Jibacoa, Habana Province. From P. J. Bermùdez.
11. Trinidad Leaseholds well Marac 1, 8965 ft., Naparima Hill formation, near base of *Globotruncana concavata* zone of Bolli (1957) (lower Santonian), coordinates N:151141 links, E:424447 links. From J. B. Saunders, Texaco Trinidad collection.

TAXONOMY

The species described in this paper are from well-preserved assemblages, and skeletal structure has been investigated in sufficient detail to form a basis for evaluating homologies between one form and another. Thus, an attempt has been made to determine critically the degrees of relationship between the various species. Difficulties arise, however, in comparing the species described here with many of those in the existing literature. The latter have generally been based on less well-preserved material, and have often been assigned rather uncritically to genera within the Haeckelian system. For these reasons most of the earlier descriptions do not contain sufficient detail to permit close comparison with the forms considered here.

It is not practical to fit some of the genera and species as defined herein into the Haeckelian system of family classification, as in many cases different features of species assigned to the same genus would require their inclusion in different families—as, for instance, *Myllocercion acineton* and *M. rhodanon*; the three feet of the latter species would place it in the family Podocyrtida, while the former, without feet, would be in the family Theocyrtida.

As a great deal more information on Mesozoic forms is necessary before an attempt can be made at establishing natural relationships on the family level, the genera described have been formally grouped only under Spumellaria and Nassellaria. Artostrobiids are described under the genus *Theocampe*, cannobotryids under the genera *Eribotrys* and ? *Bisphaerocephalina*, and stichocyrtids under *Lithocampe, Dictyomitra, Stichopilidium, Stichomitra* and *Amphipyndax*. Acanthodesmiids present in samples from Cima Hill, Moreno Gulch, and CAS loc. 39545 (their earliest known occurrence) have not been described because specimens from some of these samples are being considered by Robert M. Goll in his extensive revision of this group.

During the course of this study it has become increasingly apparent that, at least in the stichocyrtids of the Cretaceous, the Haeckelian generic definitions cannot be used meaningfully. One of the main generic distinctions of this system, presence or absence of a horn, varies within some species when the horn is small; and another, shape and degree of constriction of the last segment and its aperture, appears to be applicable for the distinction of species but not of genera. Characteristics considered important in generic distinctions are (internally) the cephalic structure s and (externally) the relative size of the cephalis and thorax, their structural relationship to each other, and the arrangement of pores. See, for example, the first three species assigned to ? *Stichomitra* (Pl. 8, figs. 1–3) which are all clearly closely related by the similarity of the cephalis and its relation to the thorax. However, according to the Haeckelian system these would be widely separated not only at the generic but even at the subfamily level.

Some morphological terms are herein used in a rather different or more restricted sense than they have been in the past, and some new terms are introduced. These are discussed below:

Cone. A wide open pylome with lamellar wall, developed in some spongy spumellarians. (Pl. 2, figs. 1*b* and 3*b*. See discussion under ?*Staurodictya fresnoensis* for details).

Differentiated large pore, sometimes prolonged. In some dicyrtids and tricyrtids with inflated terminal segment, this large segment has a differentiated large pore near the stricture separating it from the previous segment. This stricture may be the collar or lumbar one, or the junction with an enclosed thorax or cephalis. The differentiated pore is sometimes subdivided into several pores, and sometimes surrounded by a protruding rim or 'tube'. It is generally situated approximately in line with the right primary lateral spine (Pl. 2, fig. 8*a*).

Besides the asymmetry created by the differentiated large pore described above the species assigned to the genera ?*Tetraphormis* and ?*Sciadiocapsa* also show a similar asymmetry in that the portion of the flared thorax or ?abdomen associated with the right primary lateral spine tends to be broader (longer) (text-fig. 1, fig. 2*a*).

Prongs. It is well known that pores in some forms are subdivided by secondary meshwork. When this subdivision is not complete, the small spines that extend inward from the pore margin are termed 'prongs'.

Tube extending from vertical spine. In the artostrobiids, cannobotryids, and the forms described herein belonging to the genera ?*Clathrocyclas, Cornutovum, Ectonocorys,* ?*Ectonocorys,* ?*Lithomelissa,* ?*Lophophaena, Rhopalosyringium, Sciadiocapsa,* and ?*Tetraphormis* a cephalic tube extending from the vertical spine may be directed upward, laterally, or downward. When the tube extends laterally, or downward along the thoracic wall, a pair of pores or a divided single pore penetrates that part of the thoracic wall which forms part of the tube. Rarely the tube, as in *Lithomelissa heros,* is little more than an enlarged, rimmed pore. The vertical spine is attached to the proximal wall of the tube by paired branches and an extension of the vertical spine extends into, or rarely beyond, the margin of the tube. Thus, unlike younger forms, the Maestrichtian forms tend to have tubes rather than horns associated with the vertical spine.

Unless otherwise indicated, all dimensions given in the species descriptions are based on measurements of specimens from CAS loc. 1144, Marca shale (MBP), and Moreno Gulch.

Type and illustrated specimens have been deposited in the United States National Museum, Washington, D.C., and are assigned to catalogue 33, nos. 157901–158024. England finder slide number after the USNM number indicates position of specimen on slide.

1b
1a
1c
2a
A
2b
3a
A V T
3b
A
V
I
L
D
4a
4b
V
A
L
4c
5
6
8a
8b
DB
LR
A
I
L
Ax
9a
9b
10a
10b
7
Cerv
V
L
Card
A
I
D
Jug
11d
11a
l
a
h
e
g
?
11b
11c
LR
A
I
L
Ax
Card
11e

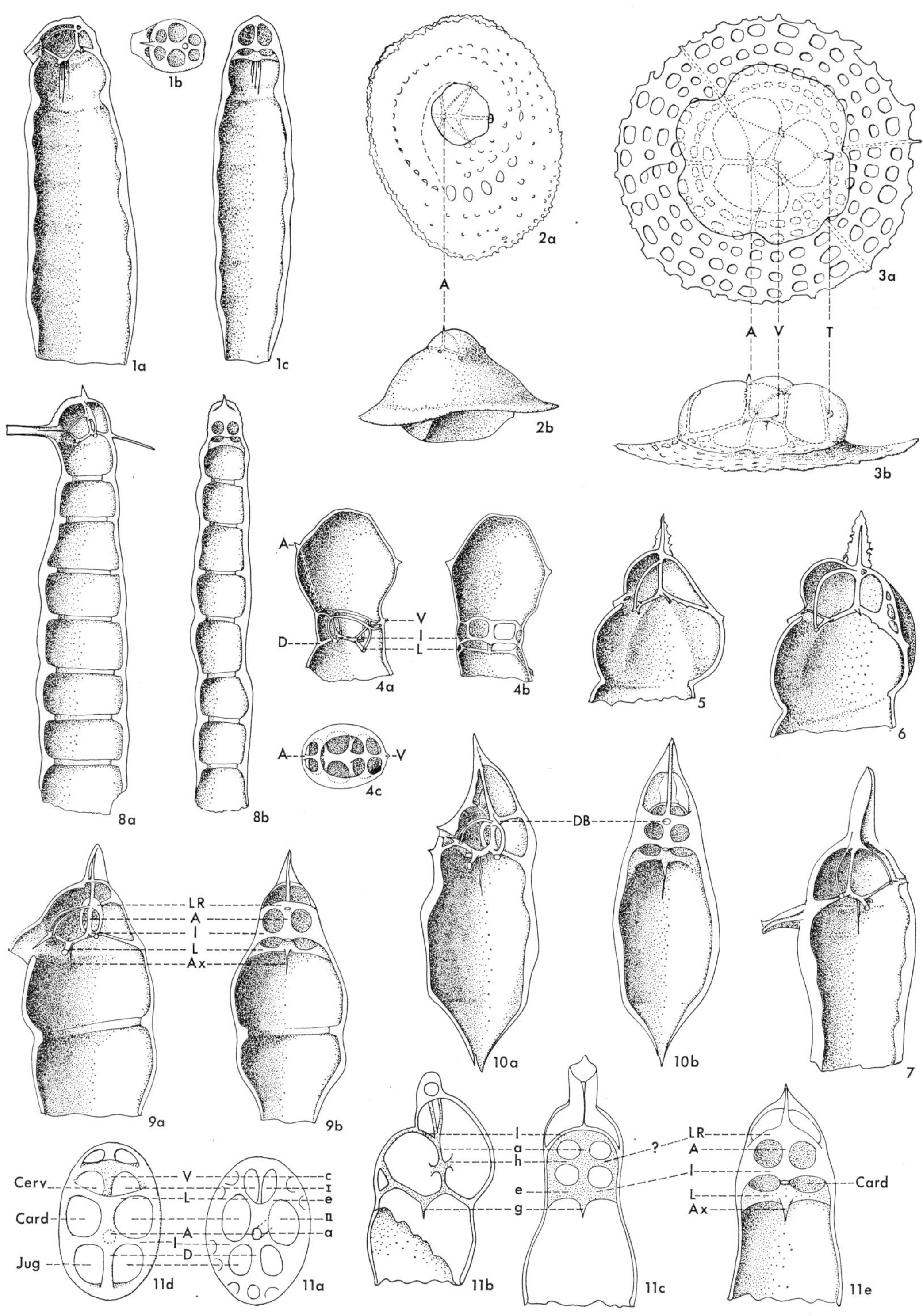

Text-fig. 1.

SYSTEMATIC DESCRIPTIONS

LEGION SPUMELLARIA

Genus *Spongosaturnalis* Campbell and Clark 1944*b*

Spongosaturnalis campbelli sp. nov.

Plate 1, figs. 5*a, b*

Description. Shell spherical to ellipsoidal, of dense, fine, spongy meshes. Ring elliptical with ridge on inner edge, which may become weak at the narrow ends and which extends across the ends of the polar spines and not on them. Ring generally bilaterally symmetrical. Three to five (generally three) smooth spines on each half ring; the terminal ones longest. Polar spines short, smooth, cannot always be distinguished when shell extends completely across ring.

Length (measured from outer edge) of long axis of ring 230–320 μ (most 250–70 μ), of short axis 125–55 μ, of terminal spines 45–120 μ, of polar spines 5–20 μ; width of shell in long axis of ring 75–130 μ, in short axis of ring 75–110 μ. Dimensions based on twenty specimens.

TEXT-FIG. 1. Details of cephalic structure. All figures drawn without pores except fig. 2*a* and 3*a, b*. Dorsal view sections through plane of apical spine and secondary lateral spines.

Abbreviations (Bütschli's key in parenthesis)

A (a)	apical spine or horn		DB	dorsal branch
V (c)	vertical spine or horn		LR (l)	lateral ridge
D	dorsal spine or wing		(h)	median branch
L (e)	primary lateral spine or wing		Card	cardinal pore
l	secondary lateral spine		Cerv	cervical pore
T	tube		Jug	jugular pore
AX (g)	axial spine			

Figs. 1*a–c. Theocampe argyris* sp. nov. *a*, section, left lateral view. *b*, collar pores, apical view. *c*, section, dorsal view.

Figs. 2*a, b.* ? *Sciadiocapsa ptesimolecis* sp. nov. *a*, apical view showing internal cephalic structures. *b*, right lateral view.

Figs. 3*a, b.* ? *Tetraphormis lobocarena* sp. nov. *a*, apical view showing internal cephalic structures (dotted lines are folds in surface of cephalis). *b*, right lateral view.

Figs. 4*a–c.* ? *Tripodocorys megalocarena* sp. nov. *a*, section, right lateral view. *b*, section, dorsal view. *c*, collar pores, apical view.

Fig. 5. *Ectonocorys deltota* sp. nov., section, left lateral view.

Fig. 6. ? *Ectonocorys hemicarena* sp. nov., section, left lateral view.

Fig. 7. ? *Lithomelissa heros* Campbell and Clark, section, right lateral view. Apical-vertical and apical-secondary lateral ridges as shown in this drawing are not constantly present.

Figs. 8*a, b.* ? *Bisphaerocephalina apimelos* sp. nov. *a*, section, left lateral view. *b*, section, dorsal view.

Figs. 9*a, b. Eribotrys despoena* sp. nov. *a*, section, left lateral view. *b*, section, dorsal view.

Figs. 10*a, b. Eribotrys litos* sp. nov. *a*, section, left lateral view. *b*, section, dorsal view.

Figs. 11*a–e. a–c, Lithobotrys geminata* Ehrenberg (Bütschli 1892, pl. 33, figs. 27*a, b, c*). *a*, collar pores, basal view (compare with 11*d*). *b*, section, left lateral view. *c*, section, dorsal view (compare with 11*e*). *d, e Eribotrys* sp. *d*, collar pores, basal view. *e*, section, dorsal view.

Discussion. This species differs from *Saturnalis ellipticus sens. lat.* Squinabol (1914) in having a ring with an inner ridge, and from *Spongosaturnalis nematodes* as described under that species. It differs from a similar species in the lower Santonian of Trinidad in that on the latter the outer edge of the ring near the polar spines is flattened so that the ring at this point has a ridge on the outer edge.

This species is named after Arthur S. Campbell who described the first comprehensive Cretaceous radiolarian fauna on the North American continent.

Localities. CAS loc. 1144, Marca shale (MBP), Moreno Gulch, Cima Hill hole III (at approximately 340 ft.) and Evitt slide AE24.

Type specimens. Holotype USNM 157901, N37/0 CAS loc. 1144; paratype USNM 157902, K37/0 Marca shale (MBP), both from upper Maestrichtian Moreno formation, Fresno County, California.

Spongosaturnalis nematodes sp. nov.

Plate 1, fig. 4

Description. Shell, never observed complete, loosely spongy; attached to elliptical ring by a series of irregular spines or threads, central one on each side sturdier than others. Ring with a ridge on inner edge which extends across the ends of the irregular spines. Five to eight (generally six or seven) spines on each half ring; terminal spine long with a median ridge, others short, smooth or rarely with a weak ridge.

Length (measured from outer edge) of long axis of ring 250–320 μ, of short axis 95–125 μ, of end spine 10–155 μ (most 45–120 μ). Dimensions based on fifteen specimens.

Discussion. This species differs from *S. campbelli* in lacking a distinct polar spine, and in having more spines on the ring with a ridge on the terminal spines and a loose spongy shell. A related form in the lower Maestrichtian–?upper Campanian CAS loc. 39545 differs in having terminal spines generally short, rarely with ridge, and short spines tending to be concentrated proximally (near position of irregular spines or threads).

Localities. CAS loc. 1144, Marca shale (MBP), Moreno Gulch, and Cima Hill hole III (at approximately 327·5 ft.).

Holotype. USNM 157903, N38/1 from upper Maestrichtian Moreno formation, CAS loc. 1144, Fresno County, California.

EXPLANATION OF PLATE 1

Figs. 1*a–f*. ? *Spongosaturnalis catadelos* sp. nov. *a*, holotype USNM 157906. *b–f*, paratypes. *b*, *c*, USNM 157908; *b*, M40/0; *c*, M39/0. *d*, USNM 157907. *e*, *f*, USNM 157908; *e*, M41/0; *f*, M42/0.
Figs. 2*a–c*. ? *Saturnalis deiropede* sp. nov. *a*, paratype USNM 157911. *b*, holotype USNM 157909. *c*, paratype USNM 157910.
Figs. 3*a*, *b*. *Spongosaturnalis spiniferus* Campbell and Clark. *a*, USNM 157905. *b*, USNM 157904.
Fig. 4. *Spongosaturnalis nematodes* sp. nov., holotype USNM 157903.
Figs. 5*a*, *b*. *Spongosaturnalis campbelli* sp. nov. *a*, holotype USNM 157901. *b*, paratype USNM 157902.
All figures × 104.

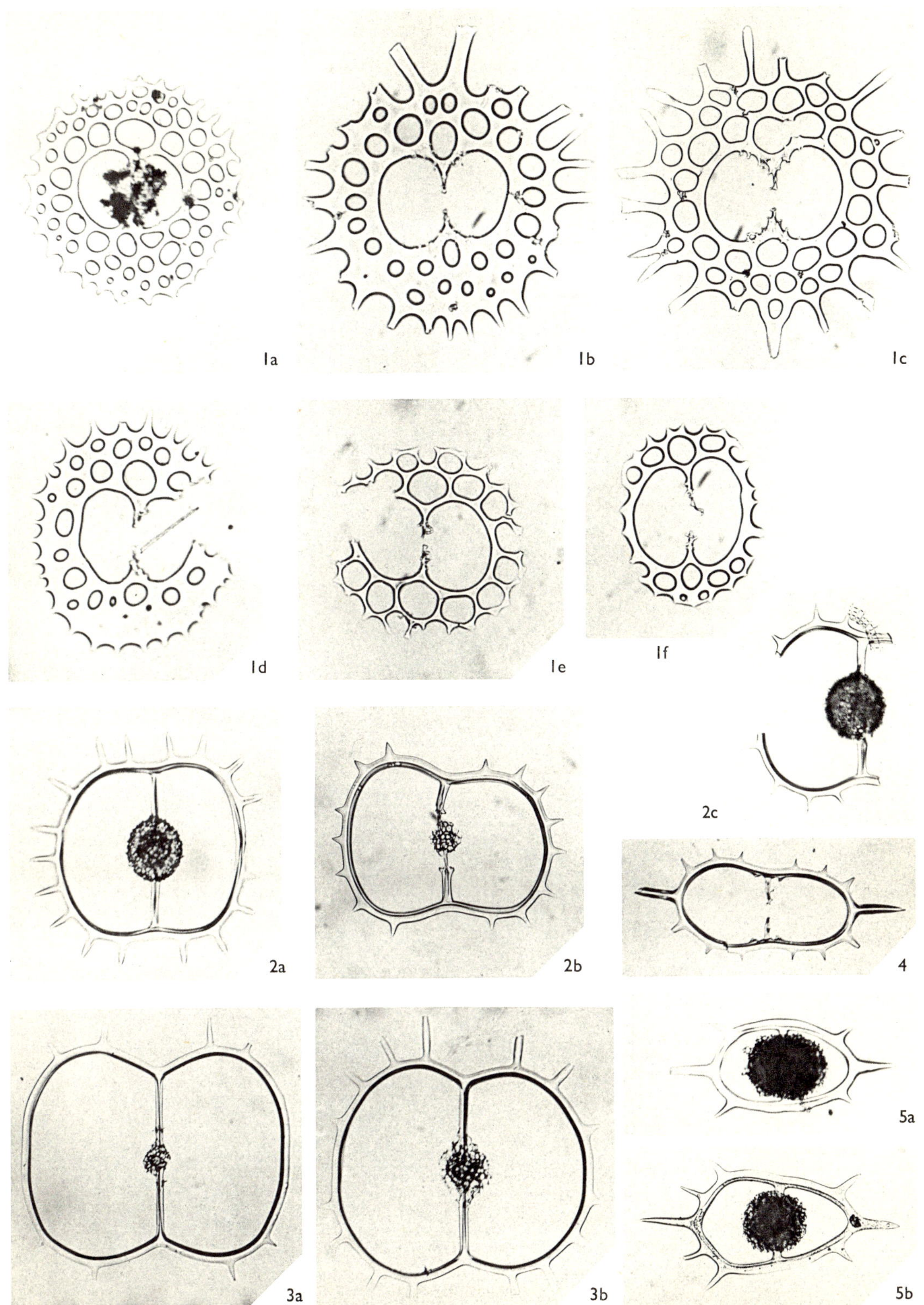

FOREMAN, Upper Maestrichtian Radiolaria of California

Spongosaturnalis spiniferus Campbell and Clark emend.

Plate 1, figs. 3*a, b*

1944*b* *Spongosaturnalis spiniferus* Campbell and Clark, p. 7, pl. 2, figs. 1–5.

Description. Shell approximately spherical, spongy, denser centrally, of very irregular meshes. Looser outer part generally not present, so that central mass is exposed. Polar spines with central ridge originate from this central area, extend and bifurcate to form a subcircular to subquadrangular ring with distinct indentation proximally. Rarely polar spines not exactly in line so that one part of ring markedly smaller than other part, and very rarely a third spine is present dividing the ring into three approximately equal parts. Ridge of polar spine also bifurcates to join ridge on or near inner edge of ring. Four to fifteen flat spines on ring, varying in size and shape from short, sharp, thornlike to longer, slender, with rounded or sharp ends. Number of spines on each half ring generally unequal; they tend to be concentrated on and near the proximal part of the ring when they are few, and distributed unevenly when more numerous.

Diameter of shell 85–110 μ, of ring (measured along polar spines) 257–97 μ, transversely 325–440 μ; length of polar spines (from outer spongy surface) 60–90 μ. Dimensions and description based on twelve specimens from CAS loc. 39545.

Discussion. Specimens from UCMP loc. A2615, the type locality, differ from specimens from the other localities mentioned below in their larger size and in having more spines on the ring. A single fragmentary specimen from Trinidad well Marac 1 differs considerably in having very much thicker polar spines and a much smaller spongy shell, and is not now included in this species.

Localities. CAS loc. 1144, Marca shale (MBP), Moreno Gulch, CAS loc. 39545, UCMP loc. A2615, Evitt slide AE24, and Lamont V–18–129.

Illustrated specimens. USNM 157904, N38/2 and USNM 157905, N38/2, both from lower Maestrichtian–? upper Campanian 'Moreno Grande' formation, CAS loc. 39545, Alameda County, California.

? *Spongosaturnalis catadelos* sp. nov.

Plate 1, figs. 1*a–f*

Description. Complete shell never seen, but believed (from fragments attached to polar spines and on ring where polar spines bifurcate) to be at least partially spongy. The development of these shell fragments or small thorns varies considerably. On some specimens they are present only on the proximal polar spines. There is also some evidence of a lattice-shell with very small pores, possibly forming the base of the spongy material. Ring circular to subcircular, broad, flat with smooth surface, perforated by numerous, rarely sparse, pores of variable size and shape; one pore, generally larger than others, at the end of each polar spine. Outer margin may be either deeply indented with long slender spines which sometimes bifurcate (generally in specimens with numerous pores) or scalloped with short, broad-based, sharp spines (generally in specimens with fewer pores).

Width (exclusive of spines) to outer margin of ring measured along polar spines 207–381 μ, transversely 183–376 μ, of inner margin measured transversely 143–257 μ; number of spines on margin 16–34. Dimensions based on fifteen specimens.

Discussion. The generic assignment of this species is doubtful because the nature of the shell is not known (although the presence of numerous small thorns on the polar spines suggests that the shell may be spongy). While even complete rings are rare, fragments are common, distinctive, and easily recognizable.

Localities. CAS loc. 1144, Marca shale (MBP), Moreno Gulch, and Cima Hill hole III (at approximately 310 and 340 ft.).

Type specimens. Holotype USNM 157906, P37/2 Marca shale (MBP); paratypes USNM 157907, N37/0 CAS loc. 1144, and USNM 157908, M39/0, M40/0, M41/0, M42/0 Moreno Gulch (3518), all from upper Maestrichtian Moreno formation, Fresno County, California.

Genus *Saturnalis* Haeckel 1881, emend. Nigrini 1967

? *Saturnalis deiropede* sp. nov.

Plate 1, figs. 2a–c

Description. Outer shell spherical to ellipsoidal, longest dimension along polar axis; pores small, subcircular, of variable size; by-spines long, slender, curved, or short. Inner shell spongy throughout, spherical to sub-spherical, with inner meshes larger and sturdier than outer ones. The whole attached to outer shell by numerous, slender, rod-like, secondary radial beams and by two opposed, sturdy, ridged primary radial beams which originate within the spongy shell and continue beyond outer shell as ridged polar spines to join the elliptical to subquadrangular spiny ring. Ring widest transversely and slightly narrowed to its smallest dimension where it meets the polar spines. The two parts frequently unequal in size and shape. The ridge, near or on its inner edge, continues across ends of polar spines. Spines on ring rarely equal in number on the two parts; they are generally flat and vary in size and shape from short thornlike to longer, slender, with sharp or rounded ends.

Discussion. Specimens from the upper Maestrichtian of Fresno County agree with specimens from the lower Maestrichtian–? upper Campanian of Alameda County except that those from the latter locality tend to have a thinner, more elliptical outer shell, longer polar spines, and more numerous spines on the ring. Although they are here considered as one species, separate dimensions are given in the following table to facilitate their comparison. In the upper Maestrichtian localities studied specimens with shell attached are rare, but fragments of the distinctive ring and polar spine are common. Unless, however, there is some indication of the presence of an outer lattice-shell, fragments cannot be positively identified, because there is another rare undescribed species in CAS loc. 39545 with similar ring and polar spines, but with a completely spongy shell. ? *Saturnalis deiropede* is distinguished from *Spongosaturnalis spiniferus* in having the ridge of the ring extending across the end of the polar spines with the ring at this point not at all or only slightly indented, and the presence of an outer lattice-shell.

There is some doubt about the generic assignment of this species because the inner shell is spongy.

Localities. CAS loc. 1144, Marca shale (MBP), Moreno Gulch, Cima Hill hole III (at approximately 340 ft.), CAS loc. 39545, and Evitt slide AE24.

Type specimens. Holotype USNM 157909, M42/1; paratype USNM 157910, Q41/2, both from upper Maestrichtian Moreno formation, Moreno Gulch (3518), Fresno County, California; paratype USNM 157911, M38/2 from lower Maestrichtian–?upper Campanian 'Moreno Grande' formation, CAS loc. 39545, Alameda County, California.

TABLE 2. Dimensions for specimens from four upper Maestrichtian localities and one lower Maestrichtian

	Upper Maestrichtian (*20 specimens*)	*Lower Maestrichtian* (*20 specimens*)
Diameter of shells:		
outer along polar axis	75–100 μ (most 90–100 μ)	75–100 μ (most 90–100 μ)
transversely	85–95 μ (most 90–5 μ)	80–95 μ (most 80–90 μ)
inner along polar axis	35–45 μ	45–50 μ
transversely	35–50 μ	45–60 μ
Length of polar spines	35–55 μ (most 45–50 μ)	50–95 μ (most 55–65 μ)
Diameter of ring (to outer margin)		
along polar spines	178–222 μ (most 200–22 μ)	200–306 μ (most 222–57 μ)
Half-diameter of ring transversely	138–85 μ (most 143–58 μ)	125–212 μ (most 138–65 μ)
Number of spines on half ring	4–12 (most 6–9)	6–15 (most 7–11)

Genus *Cenellipsis* Haeckel 1887

Cenellipsis heteroforis Campbell and Clark emend.

Plate 2, fig. 4

1944b *Cenellipsis (Cenellipsula) heteroforis* Campbell and Clark, p. 12, pl. 5, fig. 7.

Description. Shell ellipsoidal, rarely subspherical, frequently slightly flattened at one pole. Pores circular to angular, extremely variable in shape, size, and distribution. Ridges on the intervening bars, when well developed, form a rough surface; ridges sometimes more pronounced at one pole.

Length of shell 90–185 μ; diameter of shell 60–135 μ, of pores 1–10 μ. Dimensions based on twenty specimens.

Discussion. Specimens from Moreno Gulch tend to be larger than those from the other localities. This species differs from *Cenellipsis elliptica* Lipman (1952) in having larger pores, very variable in size and shape.

Localities. CAS loc. 1144, Marca shale (MBP), Moreno Gulch, CAS loc. 39545, UCMP loc. A2615, Stanford loc. PL2875, and Trinidad well Marac 1.

Illustrated specimen. USNM 157912, O38/4 from upper Maestrichtian Moreno formation, CAS loc. 1144, Fresno County, California.

Genus *Staurodictya* Haeckel 1881 emend. Kozlova and Gorbovetz 1966

? *Staurodictya fresnoensis* sp. nov.

Plate 2, figs. 1*a–e*

Description. Flat, centrally depressed, rectangular to subrectangular, spirally chambered lattice-shell with circular to subcircular pores, variable in size; and commonly four, approximately equal, three-bladed main spines, one at each corner originating near the centre. Shell structure not in a continuous spiral but increasing in steps at the points where it crosses the main spines. Chambers penetrated by numerous small radial beams, some of which continue between more than one whorl. By-spines over entire surface. While the majority of specimens have four spines, one at each corner, a fifth weaker spine is sometimes present on one side of the rectangle. Rarely there are five spines of equal size evenly distributed forming a pentagon. A large, wide open cone, originating near the centre of the shell, may be present. It is smooth with widely and irregularly spaced circular pores decreasing in size distally and a ragged margin. One or two of the main spines may be incorporated in its wall; there is no constant relation between its position and the corners of the rectangle. Specimens with cones tend to be small and generally have main spines less regularly distributed.

Breadth of rectangle 70–193 μ, length of spines 60–153 μ, diameter of cone at shell margin 75–110 μ, number of spiral whorls 3–6. Dimensions based on thirty specimens.

Discussion. Some less regular forms, excluded from the above description, co-occur in the upper Maestrichtian. This species differs from *Staurodictya* (?) *densa* Kozlova (Kozlova and Gorbovetz 1966) in that the chambers of the spiral are less embracing, and the disc therefore thinner. It would appear that this species might equally well have been assigned to the genus *Stylodictya* Ehrenberg (1847*a*); the doubt cannot be resolved until a thorough study is made of the entire group.

In many of the chambered or spongy spumellarians examined there is a pylome-like development which is here termed 'cone'. This cone arises approximately at the centre of the shell and extends with almost straight sides to form a very wide aperture at the margin or, more frequently, extends beyond the shell with a lamellar porous wall.

EXPLANATION OF PLATE 2

Figs. 1*a–e*. ? *Staurodictya fresnoensis* sp. nov. ×166. *a*, holotype USNM 157913. *b–e*, paratypes. *b*, USNM 157914; *c*, USNM 157915; *d*, USNM 157916, section along plane of spines; *e*, USNM 157917, section along lateral plane.

Figs. 2*a–c*. ? *Tripodocorys megalocarena* sp. nov. ×251. *a*, holotype USNM 157921, right lateral view. *b*, *c*, paratype USNM 157922; *b*, left lateral view; *c*, dorsal view.

Figs. 3*a*, *b*. ? *Lithelius robustus* (Campbell and Clark) ×166, USNM 157919.

Fig. 4. *Cenellipsis heteroforis* Campbell and Clark ×251, USNM 157912.

Fig. 5. ? *Cryptocapsa axios* sp. nov. ×251, holotype USNM 157923.

Fig. 6. ? *Cryptocapsa asymmetros* sp. nov. ×251, holotype USNM 157924.

Figs. 7*a*, *b*. *Pseudoaulophacus* sp., USNM 157918. *a*, ×166; *b*, surface detail ×415.

Figs. 8*a–c*. *Gongylothorax verbeeki* (Tan Sin Hok). *a*, USNM 157925, apical view showing differentiated large pore associated with right primary lateral spine ×251. *b*, *c*, USNM 157926; *b*, surface detail ×415; *c*, ×251.

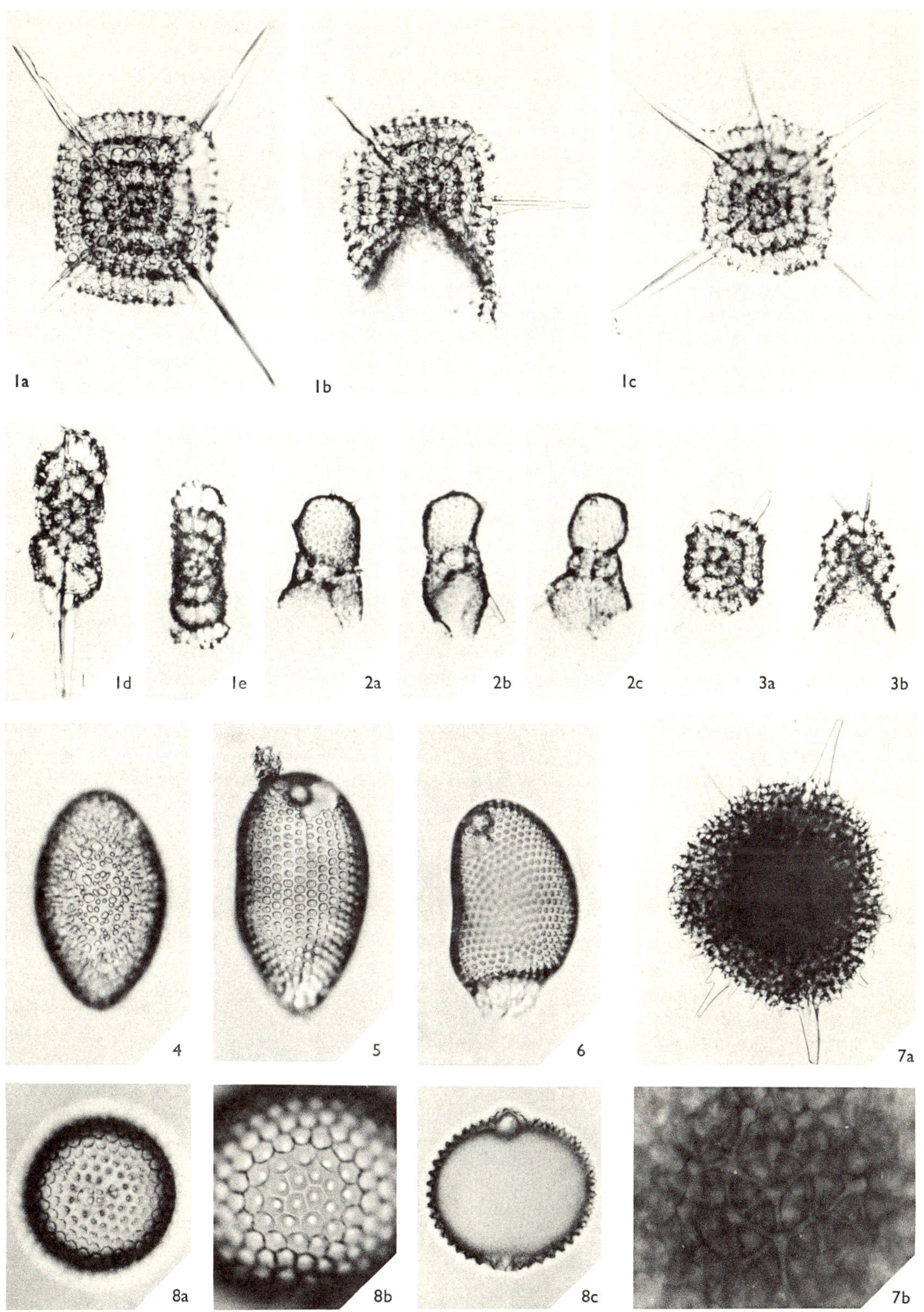

FOREMAN, Upper Maestrichtian Radiolaria of California

According to the thickness of the shell the cone may be elliptical or circular in cross-section. Up to three cones have been observed on one specimen. However, when two or three cones are present so little of the original shell material remains that it is difficult to determine their species relationships. This development of a wide cone has been observed in specimens from the ?Aptian Thirindine shale through the upper Maestrichtian and in ? Paleocene of Cima Hill hole III at approximately 280·5 ft. It may be that it extends throughout the Mesozoic, but no samples older than the Thirindine shale have been examined. The cone described above for some species in the Cretaceous differs from the pylome described in Tertiary and Recent species in being much broader and frequently extending beyond the shell margin with a lamellar porous wall. According to Riedel (personal communication) this wide cone structure has not been observed in Eocene and younger radiolarians.

Hülsemann (1963), as a result of her study of Recent *Spongotrochis* (?) *glacialis*, concluded that there is no justification for generic separation of forms on the basis of presence or absence of pylome (as by Dreyer 1889). She considered also that the species with a pylome for which the genus *Schizodiscus* Dogiel (Dogiel and Reshetnyak 1952) was established probably represent pylomatic forms of species which also have forms without pylomes, and that therefore *Schizodiscus* is not a useful genus. The author agrees that as in the case of the Tertiary forms with or without pylome, Mesozoic forms with cones are simply variants of species without cones. Two Mesozoic species with cones have been reported earlier. Campbell and Clark (1944*b*) described a form with a cone which they interpreted as a nassellarian *Lithomelissa* (*Micromelissa*) *robusta* (see ? *Lithelius robustus*, below), and Kozlova (Kozlova and Gorbovetz 1966) described *Spongopyle insolita*.

Localities. CAS loc. 1144, Marca shale (MBP), Moreno Gulch, Cima Hill hole III (at approximately 285, 306·5, 310, 313·5, 320, 323·5, 327·5, 330, 340, 356·5, and 360 ft.), and Lamont V–18–129.

Type specimens. Holotype USNM 157913, M38/4; paratypes USNM 157914, M38/2, USNM 157915, N38/2, USNM 157916, P38/2, USNM 157917, P37/4, all from upper Maestrichtian Moreno formation, Moreno Gulch (3518), Fresno County, California.

Genus *Pseudoaulophacus* Pessagno 1963

Plate 2, figs. 7*a*, *b*

Rare specimens belonging to this genus were found in CAS loc. 1144, Marca shale (MBP) and Moreno Gulch, but not in sufficient numbers to provide an adequate basis for species descriptions. It occurs commonly in Cuba B191 and has been seen in Trinidad well Marac 1.

Illustrated specimen. USNM 157918, M38/2 from upper Maestrichtian Moreno formation, CAS loc. 1144, Fresno County, California.

Genus *Lithelius* Haeckel 1860

? *Lithelius robustus* (Campbell and Clark)

Plate 2, figs. 3*a*, *b*

1944*b Lithomelissa* (*Micromelissa*) *robusta* Campbell and Clark, p. 26, pl. 7, fig. 8.

Description. Spherical to subspherical lattice-shell of approximately three spiral whorls, with or without a cone, and with eight three-bladed main spines. Whorls connected by numerous smooth secondary radial beams, and the outer whorls, at least, by the three-bladed beams which extend inward from the main spines. These main spines approximately equal on individual specimens, but varying considerably in length among specimens. Pores are circular to subcircular and the surface is covered with by-spines of which some continue from the secondary radial beams and others originate at the shell surface. Cone, when present, originates near the centre and extends beyond outer margin of shell as a smooth lamellar porous wall; pores circular to subcircular, irregularly arranged, smaller and more widely spaced distally; margin ragged. One or two main spines may be incorporated in wall of cone.

Diameter of shell 70–100 μ, greatest width of cone 75–120 μ, length of spines 35–85 μ. Dimensions based on twenty-five specimens.

Discussion. This species was described by Campbell and Clark (1944b) who considered it a nassellarian, and interpreted the spiral sphere as the cephalis and the cone as the thorax. They were not certain of this interpretation and suggested that 'the species belongs to some other than the assigned genus'. Its assignment here doubtfully to *Lithelius* is not meant to imply that there is any strong indication of its real relationship to the type species of that genus, but it seems inadvisable to propose a new genus until its relationships are better understood.

Localities. CAS loc. 1144, Marca shale (MBP), Moreno Gulch, Cima Hill hole III (at approximately 306·5, 310, 320, 340, and 356·5 ft.), UCMP loc. A2615, CAS loc. 39545, and Stanford loc. PL2875.

Type specimens. Holotype USNM 157919, M39/0; paratype USNM 157919, M39/0, both from upper Maestrichtian Moreno formation, Marca shale (MBP), Fresno County, California.

LEGION NASSELLARIA

Genus Archicorys Haeckel 1881

Remarks. No internal structures have been observed in the shell, and therefore the assignment of the species described below to the nassellarian genus *Archicorys* is doubtful. However, there is not any indication, in the description of the type species of that genus, of the presence or absence of internal structure, and the doubt regarding the homologies of its shell can be resolved only by investigation of topotypic material.

Superficially, the shell bears some resemblance also to *Lithocarpium pyriformis* Stöhr (1880), but the latter is described as not having pores. A search through radiolarian assemblages from three localities of the Sicilian Miocene did not reveal any specimens of this species, which Stöhr described as very rare.

? Archicorys allodarpe sp. nov.

Plate 3, fig. 4

Description. Shell an ellipsoid, at one pole a spine at the other a tube, all three very variable in length and width. The ellipsoid is smooth except for rare short thorns on the area around the tube. Pores are circular to subcircular, small near the spine and larger near the tube, without regular arrangement. The spine is approximately circular in cross-section, smooth or ridged, in some specimens slightly bifurcated. The tube has a few irregular pores or is poreless, occasionally with some thorns, its margin with a few irregular teeth.

Length of ellipsoid 65–200 μ, of spine 10–40 μ, of tube 15–45 μ; diameter of ellipsoid 50–100 μ, of its pores 1–8 μ, of spine near base 5–15 μ, of tube 12–17 μ. Dimensions based on thirteen specimens.

Discussion. One specimen, not now included in this species, was observed in CAS loc. 39545 which differed in its greater size (length of ellipsoid 272 μ, width 118 μ) and in having a few small spines around the polar spine.

Localities. CAS loc. 1144 and Moreno Gulch.

Holotype. USNM 157920, M38/0 from upper Maestrichtian Moreno formation, CAS loc. 1144, Fresno County, California.

Genus *Tripodocorys* Haeckel 1881

Remarks. Tripodocorys is here used as a generic name as it was originally by Haeckel (1881) and Rüst (1885), and not as a subgenus of *Tripilidium* as in later works. The generic assignment of ? *T. megalocarena* remains doubtful because of the presence of a second segment and because the internal structure of the type species *T. fischeri* Rüst (1885) is not known.

? *Tripodocorys megalocarena* sp. nov.

Plate 2, figs. 2*a–c*; text-fig. 1, figs. 4*a–c*

Description. Shell small, of two segments; cephalis relatively large and thorax small, with three wings and ragged distal margin. Internally, cephalis divided into two parts, a large globular upper and a cylindrical neck-like lower, by a ring which is joined to and completely encircles the cephalic wall. It is joined to the apical and vertical spines by a pair of lateral branches on each spine. The vertical spine penetrates the cephalic wall at the level of this ring, and the apical spine penetrates the dorsal face of the cephalis; therefore there appear to be five pores at the level of the ring when viewed apically (text-fig. 1, fig. 4*c*). The collar structure itself has only five collar pores, the cervical pore not being divided by the vertical spine at this level. Paired, weak branches extend upward along the cephalic wall from the primary and secondary lateral spines to join the ring at the same points as the lateral branches extending from the apical and vertical spines. Externally the vertical spine extends almost horizontally at the level of the ring as a small, slender, short horn; the apical spine is incorporated in the shell wall and protrudes from the dorsal face of the cephalis as a slender, sharp horn. Generally there are other similar horns, up to four in number on the cephalis, either randomly arranged with one usually at the apex or four horns, including the dorsal, all at approximately the same level and equidistant. Small thorns or by-spines may also be present over all. Pores are circular to subcircular, small or lacking entirely apically, gradually increasing in size to the level of the ring where there is a distinct stricture and a slight change in the character of the pores. The three thoracic wings extending from the primary lateral and dorsal spines are incorporated in the shell wall proximally. Distally they extend as long, slender, smooth, downward, and outward directed spines. Thoracic pores are circular to sub-circular, tending to be larger and more irregularly arranged than those of the cephalis. The thorax itself is approximately cylindrical in shape with some constrictions distally.

Length of total cephalis 45–75 μ (of upper part 35–60 μ, of lower part 10–15 μ), of longest thorax 70 μ, of longest cephalic horn 40 μ, of longest broken wing 65 μ; width of cephalis 35–58 μ. Dimensions based on fifteen specimens.

Discussion. Superficially this species appears similar to *Lithomelissa thoracites* Haeckel (1862). Its internal structure, however, clearly distinguishes it from this and other related species with a large globose cephalis in which the internal structure is known.

Localities. CAS loc. 1144, Marca shale (MBP), and Moreno Gulch, Cima Hill hole III (at approximately 320 ft.), and CAS loc. 39545.

Type specimens. Holotype USNM 157921, K38/0; paratype USNM 157922, K38/0, both from upper Maestrichtian Moreno formation, Moreno Gulch (3517), Fresno County, California.

Genus *Cryptocapsa* Haeckel 1881

Remarks. The most appropriate generic assignment of the species below and *C. asymmetros* is difficult to determine for two reasons. Firstly, by the description of the type species of *Cryptocapsa* (*Cryptocapsa tricyclia* Rüst 1885, p. 307) is inadequate, and secondly, by the actual relationships of di- and tricyrtid forms with a small cephalis completely enclosed in the thorax cannot be determined until a comprehensive study is made of the group.

? *Cryptocapsa axios* sp. nov.

Plate 2, fig. 5

cf. 1951 *Cenellipsis heteroformis* Galavis, pl. 1, fig. 1; non *C. heteroforis* Campbell and Clark 1944b, p. 12, pl. 5, fig. 7.

Description. Smooth, modified ellipsoidal shell of two segments with small poreless cephalis completely hidden in thoracic cavity. Cephalis attached to thoracic wall at apex by short threads. At its base are four collar pores from which vertical, dorsal, and primary lateral spines extend into the thoracic cavity toward or to the thoracic wall. Thorax flattened apically, proximal half hemispherical, distal half inverted conical, with or without small circular aperture. Small circular pores uniform over all, except distally where they tend to be larger. Diameter of pores on inner surface frequently smaller than on outer surface. Pores arranged quincuncially in longitudinal rows, thirty-eight to fifty per circumference.

Length of thorax 135–170 μ, of cephalis 12–16 μ; diameter of thorax 85–120 μ, of aperture 7–12 μ, of pores on outside surface 4–7 μ. Dimensions based on twelve specimens.

Discussion. This species is distinguished by the small spherical completely enclosed cephalis, the regular longitudinal alignment of thoracic pores, and the strong tendency to be closed basally. Of the twelve specimens examined, only one lacked an aperture. In this study, no other species has been observed in which the presence or absence of an aperture was not a consistent feature.

Galavis (1951) illustrated a similar form from the Venezuelan Cretaceous under the name '*Cenellipsis heteroformis*', apparently a typographical error distorting '*C. heteroforis*'. There is probably a misinterpretation involved as well, since the Venezuelan form appears to be the dicyrtid nassellarian described here (rather than a spumellarian).

Localities. CAS loc. 1144, Marca shale (MBP), Moreno Gulch, and CAS loc. 39545.

Holotype. USNM 157923, P38/0 from upper Maestrichtian Moreno formation, CAS loc. 1144, Fresno County, California.

? *Cryptocapsa asymmetros* sp. nov.

Plate 2, fig. 6

Description. Shell of three segments: small completely hidden cephalis; large ellipsoidal thorax; and small, inversely conical, fragile abdomen. The poreless cephalis, situated off-centre, has four collar pores. Primary lateral and dorsal spines extend into the thoracic cavity, sometimes branching as they join the thoracic wall. Thorax smooth, asymmetric, flattened along that side where the cephalis is attached, with pores over all except at the point where the cephalis joins the thoracic wall. Circular pores in most specimens uniform in size and regularly quincuncially arranged; in a few specimens, some variation in size and spacing interrupts this regular pattern. A few specimens have a spine variably developed (up to 15 μ) on the thorax in line with the vertical spine of the cephalis. Distally, the thorax constricts abruptly so that the lumbar stricture is only about half the diameter of the thorax at its widest point. The abdomen has wall thinner, and pores slightly larger and more irregular in size, shape, and distribution than those of the thorax. Rare complete specimens have a distal margin of small irregular teeth.

Length of cephalis 15–20 μ, of thorax 95–140 μ; diameter of thorax 65–110 μ, of lumbar stricture 25–60 μ, of pores of thorax 2–4 μ, of pores of abdomen 2–5 μ. Dimensions based on twenty specimens.

Discussion. It seems quite possible that this form with a small fragile abdomen is closely related to dicyrtid forms such as ?*C. axios*. It is distinguished from the latter by the presence of a third segment and the eccentricity of the cephalis. (See remarks under *Cryptocapsa* for discussion of generic assignment.)

Localities. CAS loc. 1144, Marca shale (MBP), and Moreno Gulch.

Holotype. USNM 157924, M33/4 from upper Maestrichtian Moreno formation, Moreno Gulch (3518), Fresno County, California.

Genus *Gongylothorax* gen. nov.

Type species. G. *verbeeki* (Tan Sin Hok).

Diagnosis. Dicyrtid forms with large inflated spherical or subspherical thorax, the latter with a distinct relatively large pore or tube near its junction with the cephalis. Cephalis may or may not be partly depressed in thoracic cavity, thorax with or without aperture.

Stylocapsa Principi (1909) differs in that it has a large well-developed spine, and probably no large pore or tube.

Etymology. The name is derived from the Greek *gongylos*, spherical and *thorax* (masculine).

Gongylothorax verbeeki (Tan Sin Hok)

Plate 2, figs. 8*a–c*

1927 *Dicolocapsa verbeeki* Tan Sin Hok, p. 44, pl. 8, figs. 40, 41.

Description. Shell of two segments; small poreless cephalis with four collar pores, and large, slightly flattened, spherical thorax with small circular aperture. Position of cephalis in relation to thorax varies. In most specimens the distal half extends into the thoracic cavity and proximally is enclosed in the thoracic wall. In some specimens, however, only a very small part of the cephalis is hidden in the thoracic cavity, the degree to which the cephalis is enclosed depending partly on the thickness of the thoracic wall. Its surface is rough. Dorsal and primary lateral spines extend into the thoracic cavity. The thorax has circular pores of uniform size set in angular frames and its aperture has a smooth rim, which in most cases is flush with the thoracic surface, but rarely protrudes. A large pore or tube is located immediately below the upper margin of the thorax; in the majority of specimens it is associated with the right primary lateral spine, in a few it is situated between the right primary lateral and the vertical spine, and rarely it lies between the vertical and left primary lateral spine.

Length 80–130 μ, of cephalis 15–25 μ (most 15–20 μ), of thorax 75–125 μ (most 75–95 μ); diameter of thorax 75–140 μ, of pores 1–5 μ, of frames 6–10 μ, of aperture 10–15 μ. Dimensions based on twenty specimens.

Discussion. Although the diameter of the pores is uniform on each specimen, there is considerable variation in pore size among specimens. No differentiation into species can be made on this feature, however, because all intermediate sizes are found.

Examination of topotypic material from Tan Sin Hok's loc. 150 reveals the presence of a tube on the thorax differing only slightly from the one described above in that the tube of the Rotti form has three pores at its junction with the thorax, whereas in the California form there is only one large pore. This is not considered sufficient evidence for separating these two forms.

Gongylothorax verbeeki differs from *Dicolocapsa cephalocrypta* Tan Sin Hok (1927) in the latter having no angular frames around the pores and from *D. exquisita* Tan Sin Hok (1927) by the basally truncated ellipsoidal shape of the thorax of the latter.

Localities. CAS loc. 1144, Marca shale (MBP), Moreno Gulch, Cima Hill hole III (at approximately 356·5 ft.), UCMP loc. A2615, Stanford loc. PL2875, Evitt slides AE26, and Cuba B191.

Illustrated specimens. USNM 157925, L35/2, USNM 157926, Q39/4, both from upper Maestrichtian Moreno formation, CAS loc. 1144, Fresno County, California.

Genus *Cornutella* Ehrenberg 1838 emend. Nigrini 1967

Cornutella californica Campbell and Clark emend.

Plate 3, figs. 1*a*–*c*

1944*b* *Cornutella californica* Campbell and Clark, p. 22, 23, pl. 7, figs. 33, 34, 42, 43.
?1898 *Cornutella tenuis* Rüst, p. 40, pl. 13, fig. 2.

Description. Shell of two segments; conical, narrow or basally flared, circular or elliptical in transverse section. Cephalis very small, poreless, defined only by a slight indentation of the shell wall but without a distinct collar stricture. It bears a long, slender, straight or wavy apical horn, smooth except for the tip which may be slightly roughened; its length and width very variable. Rare specimens also have a short, sharp, vertical horn. On the thorax most specimens have a short, sharp, upward directed spine about 10–30 μ below the base of the cephalis. Thorax poreless proximally, then with circular to elliptical pores, increasing in size and sometimes becoming angular distally. Pores are arranged quincuncially in vertical rows, eight to eighteen per circumference, except for the distal part of some specimens which have pores arranged irregularly; distal margin always ragged. Proximal and median pores are, in most cases, subdivided by a very delicate meshwork. Surface rarely smooth, either slightly roughened with blunt spines proximally or with sharper spines over all except the most distal portion; when well-developed these spines branch and join to form rounded arches. Contour of shell wall generally changes at point where the pores are less regularly arranged, and weak, internal, irregular, diagonal strictures may sometimes be distinguished there. These more prominent on specimens from the lower Maestrichtian than from the upper Maestrichtian.

Length of cephalis 3–10 μ (most 4–6 μ), longest (broken?) thorax 232 μ, of ten complete apical horns 50–350 μ (most 125–60 μ); greatest width of thorax of specimens with circular cross-section 45–75 μ, with elliptical cross-section 70–115 μ, of horn near base 4–12 μ. Dimensions based on forty specimens measured not only from the Fresno County localities but also from CAS loc. 39545, Lamont V–18–129, and Trinidad well Marac 1.

Discussion. This species has been redescribed on the basis of material from several widely separated localities including CAS loc. 39545 collected near UCMP loc. A2615, the type locality for *C. californica*. Specimens examined from Campbell and Clark's type material agree well with this emended description. *C. californica* var. *brevis* could not be distinguished as a separate taxon. *C. tenuis* Rüst 1898 may be con-specific but the drawing is so stylized and the description so brief that no decision can be made until topotypic material is available.

C. californica may be distinguished from the Recent *C. profunda sens. lat.* (Riedel 1958, Nigrini 1967) by its generally larger, stouter horn. Rare specimens with a shorter horn that fall in the size range for *C. profunda* may be distinguished by a combination of any of the following features: proximally subdivided pores or pores with prongs on their margin, surface roughened by tiny spines, and generally broader wide open thorax. These features also appear to distinguish this species from *C. clathrata* Ehrenberg (1844) from the Miocene of Caltanisetta, and *Sethoconus subtilis* Carnevale (1908) from ? Miocene of Bergonzano, Reggio Emilia.

Localities. CAS loc. 1144, Marca shale (MBP), Moreno Gulch, Cima Hill hole III (at approximately 280·5 ft.), UCMP loc. A2615, CAS loc. 39545, Stanford loc. PL2875, Evitt slide AE28, Lamont V–18–129, and Trinidad well Marac 1.

Illustrated specimens. USNM 157927, O39/0, USNM 157928, L37/3, L39/0 all from upper Maestrichtian Moreno formation, Moreno Gulch (3518), Fresno County, California.

Genus *Lophophaena* Ehrenberg 1847*b*

Remarks. The basal closure of the shell of the species below is not considered sufficient reason to exclude this species from the genus *Lophophaena*, but the generic assignment of it and the next species remains uncertain because of the inadequacy of the description of the type species *L. galea* Ehrenberg (1854, p. 245).

? *Lophophaena glaucidium* sp. nov.

Plate 3, figs. 6*a, b*

Description. Shell of two segments; a hemispherical cephalis, and a subcylindrical, sometimes medianly expanded thorax; both segments elliptical in transverse section with the largest dimensions in the sagittal plane. The cephalis has small pores, irregular in size, widely and randomly spaced, circular to subcircular. It bears an apical horn and a smaller horn which extends from an internal dorsal branch of the apical spine. A small tube, difficult to distinguish, is associated with a spine extending outward from the vertical spine, and immediately above or slightly laterally to the vertical spine are one or more outward and upward directed spines. Besides these a few random thorns or spines may also be present. The collar stricture is marked by no or only a slight indentation. Internally there are six large collar pores and an axial spine. The two cervical pores are subdivided by paired branches of the vertical spine. Very weak paired branches, sometimes one lacking, extend from the apical spine along the cephalic wall to the branches of the vertical spine. The secondary lateral spines extend as small wings at the collar stricture, and the dorsal and primary lateral spines give rise to small wings on the thorax. The thorax has circular to elliptical, irregularly arranged, widely spaced pores

---

EXPLANATION OF PLATE 3

Figs. 1*a–c. Cornutella californica* Campbell and Clark. *a,* USNM 157927. *b, c,* USNM 157928; *b,* L37/3; *c,* L39/0.

Figs. 2*a–c.* ? *Lithomelissa hoplites* sp. nov. *a, b,* holotype USNM 157936; *a,* left lateral view; *b,* surface detail. *c,* paratype USNM 157937, left lateral view.

Figs. 3*a–c.* ? *Lophophaena polycyrtis* (Campbell and Clark). *a, b,* USNM 157932; *a,* right lateral view; *b,* ventral view. *c,* USNM 157931, dorsal view.

Fig. 4. *Archicorys allodarpe* sp. nov., USNM 157920.

Figs. 5*a, b.* ? *Lithomelissa heros* Campbell and Clark. *a,* USNM 157934, dorsal view. *b,* USNM 157935, left lateral view.

Figs. 6*a, b.* ? *Lophophaena glaucidium* sp. nov. *a,* holotype USNM 157929, left lateral view. *b,* paratype USNM 157930, right lateral view.

Fig. 7. *Acidnomelos apteron* sp. nov., holotype USNM 157933, left lateral view.

All figures ×251 except fig. 2*b,* ×416.

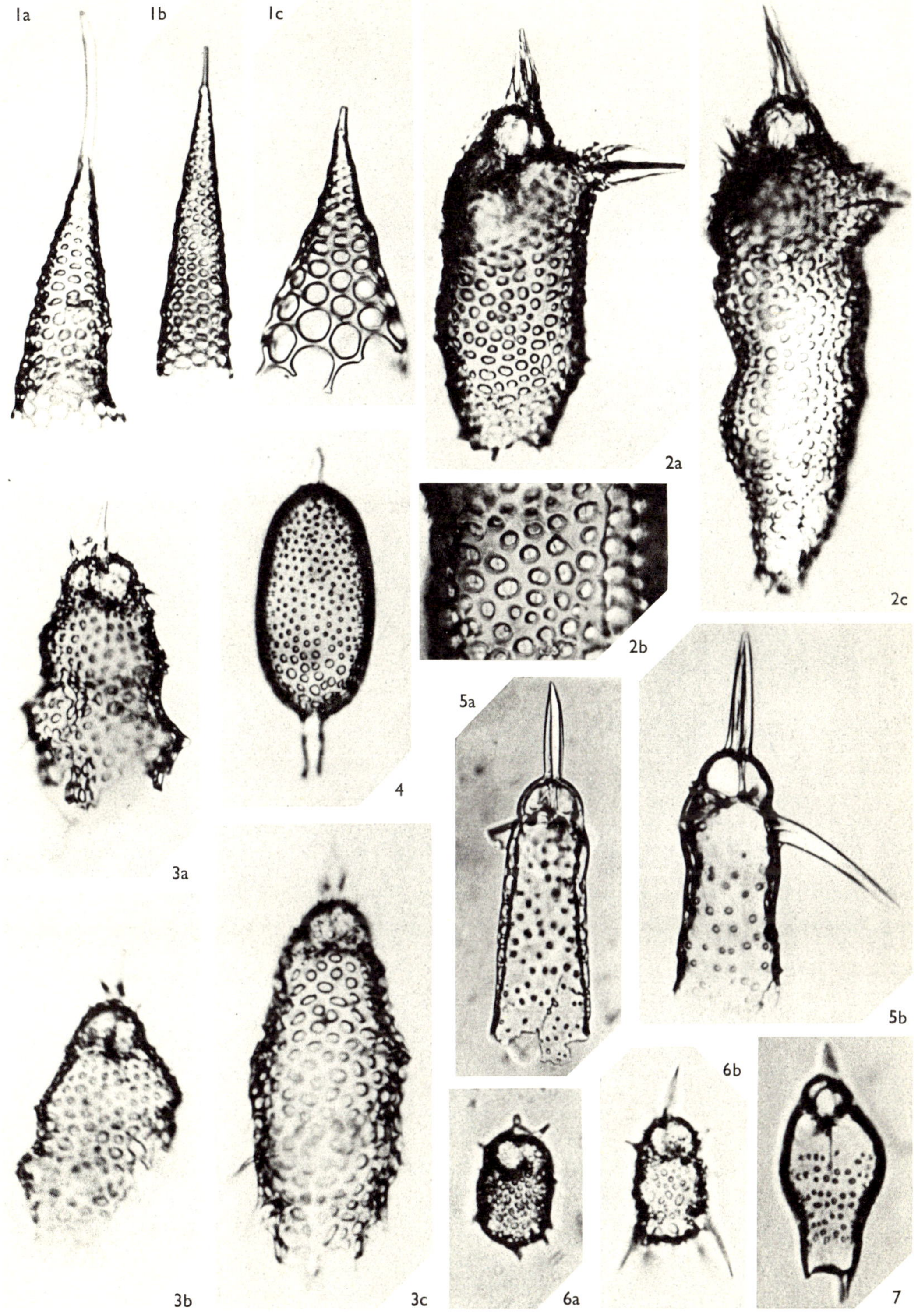

FOREMAN, Upper Maestrichtian Radiolaria of California

larger than those of the cephalis. Basally it constricts sharply, frequently to form an almost flat closed base, with large-pored mesh centrally from which depends a spine. One or more surrounding lesser spines may be present. Sturdy sharp spines (feet?) four to perhaps seven in number, originate just above the point where the thorax constricts and frequently extend below the lower margin of the thorax. Some specimens have other random, downward directed spines on the thorax.

Length of cephalis 28–35 μ, of thorax 35–65 μ; width in sagittal plane of cephalis 38–50 μ, of thorax 45–60 μ. Dimensions based on ten specimens.

Discussion. Incomplete specimens with the basal portion broken off are common; they are frequently slightly larger than the complete specimens described above and it is not entirely certain whether they belong to the same species.

Localities. CAS loc. 1144 and Moreno Gulch.

Type specimens. Holotype USNM 157929, O37/4 Moreno Gulch (3517); paratype USNM 157930, C30/4 CAS loc. 1144, both from upper Maestrichtian Moreno shale, Fresno County, California.

? *Lophophaena polycyrtis* (Campbell and Clark) emend.

Plate 3, figs. 3*a–c*

1944*b* *Sethoconus* (*Phlebarachnium*) *polycyrtis* Campbell and Clark p. 27, pl. 7, figs. 39, 40, 45, 48, 50, 51.
1944*b* *Lithomelissa* (*Sethomelissa*) *armata* Campbell and Clark, p. 26, pl. 7, figs. 44, 47.

Description. Shell of two segments. Cephalis hemispherical, spiny, with small circular to subcircular pores of variable size over all, sometimes with prongs on their margins. Six large collar pores and axial spine are present, and short spines or thorns (wings) extend from the dorsal, primary and secondary lateral spines. The apical spine branches internally below the apex of the cephalis to form, externally, a cluster of secondary spines around the main apical horn. These secondary spines variably developed from a few thorns to sturdy spines fused together basally and joined distally by horizontal bars. A short horizontal tube, frequently obscured by spines, extends from the vertical spine. Collar stricture marked by no or only a slight indentation. Thorax flared generally along the sagittal plane, with irregular expansions and constrictions more pronounced in some specimens forming a very uneven surface, distally constricting. The dorsal side tends to be straighter than the ventral side. Downward directed spines or thorns are present over all, but are more numerous distally. Pores are widely spaced, very variable in size and shape, proximally with prongs, or with prongs joining to subdivide the pores. Distal margin ragged.

Length of cephalis 30–5 μ, of longest thorax 150 μ; width in sagittal plane of cephalis 50–65 μ, of thorax 80–110 μ; width in lateral plane of cephalis 40–55 μ, of thorax 60–100 μ. Dimensions based on fifteen specimens.

Discussion. One specimen was found in CAS loc. 39545 alike in every respect except that the thorax was basally closed with a slender spine depended centrally as in ? *L. glaucidium.*

Specimens on the type slides nos. 34528, 34529 for *Sethoconus* (*Phlebarachnium*) *polycyrtis* and 34538, 34539 for *Lithomelissa* (*Sethomelissa*) *armata* have moved and it is not possible to identify confidently any individual specimens as those illustrated by Campbell and Clark. However, numerous specimens are available for study and it does not seem at all possible to separate them into two species, even though Campbell and Clark (1944*b*, p. 26) said '*L. armata*, n. sp., is so distinct and different from other species that it could scarcely be confused with any of them.' *S. polycyrtis* is chosen as the specific name because its original description and illustrations appear more typical in relation to specimens studied from other localities.

Localities. CAS loc. 1144, Marca shale (MBP), Moreno Gulch, Cima Hill hole III (at approximately 340 and 356·5 ft.), CAS loc. 39545, UCMP loc. A2615, Stanford loc. PL2875, Evitt slide AE26, and Lamont V–18–129.

Illustrated specimens. USNM 157931, R33/3 Moreno Gulch (3518) and USNM 157932, L41/3 CAS loc. 1144, both from upper Maestrichtian Moreno formation, Fresno County, California.

Genus *Acidnomelos* gen. nov.

Type species. *A. apteron* sp. nov.

Diagnosis. Dicyrtid forms with large cephalis; thorax prolonged as three lamellar pointed feet of similar structure as the thoracic wall.

This genus is distinguished from *Sethocyrtis* Haeckel (1887) (see Riedel 1957 and Nigrini 1967), and *Sethocorys* Haeckel (1887) by the presence of three terminal feet and the apparent different cephalic structure.

Etymology. The name is derived from the Greek *akidnos*, weak and *melos*, limb (neuter).

Acidnomelos apteron sp. nov.

Plate 3, fig. 7

Description. Shell of two segments, spindle-shaped. Cephalis approximately hemispherical, hyaline, poreless, with tiny papillae. It bears a sturdy, smooth apical horn, broad in sagittal plane and narrow laterally; and a small tube that protrudes only slightly or not at all, and is in line with the vertical spine. Internally, variably developed paired ridges extend along the cephalic wall from the apical spine to the side of the vertical spine. Paired ridges from the apical spine to the secondary lateral spines can sometimes be distinguished. Six collar pores and an axial spine are present. Externally the collar stricture is marked by no or only a very slight change of contour. Most specimens have a very small dorsal wing, and rarely tiny thorns protrude from the primary lateral spines. Pores are small, circular to elliptical, variable in size, widely and irregularly spaced, and the surface sometimes has tiny papillae over all. The thorax is slightly flattened, expanded medianly in the sagittal plane, and distally constricted with three short, lamellar, broad-based, pointed feet (projections of the thoracic wall) at the apertural margin.

Length of cephalis 25–30 μ, of thorax 75–110 μ, of apical horn 25–45 μ; diameter of cephalis 40–5 μ, of thorax in sagittal plane 60–72 μ, in lateral plane 48–58 μ, of horn in sagittal plane 10–15 μ, in lateral plane 5–7 μ. Dimensions based on fifteen specimens.

Localities. CAS loc. 1144, Marca shale (MBP) and Moreno Gulch.

Holotype. USNM 157933, M37/4 from upper Maestrichtian Moreno formation, Marca shale (MBP), Fresno County, California.

Genus *Lithomelissa* Ehrenberg 1847*b*

Remarks. The generic assignment of the following three species remains doubtful because of inadequacy of the description of the type species of *Lithomelissa* (*L. tartari* Ehrenberg 1854, p. 245).

? *Lithomelissa heros* Campbell and Clark emend.

Plate 3, figs. 5*a*, *b*; text-fig. 1, fig. 7

1944*b* *Lithomelissa heros* Campbell and Clark, p. 25, pl. 7, fig. 23.

Description. Shell of two segments. Cephalis hemispherical, hyaline, smooth or with tiny papillae, occasionally with a few small, circular pores dorsally, rarely over all. It bears a sturdy, ridged, apical horn, and a small tube that rarely protrudes is in line with the vertical spine. In some specimens variably developed paired ridges extend along the inner wall from the apical spine to sides of the vertical spine, and from the apical spine to the secondary lateral spines. Six large collar pores and an axial spine are present. Externally, collar stricture marked by no, or only a very slight change in contour. The thorax, basically a gently tapering cylinder, expands slightly proximally where three, ridged or smooth, sturdy dorsal and primary lateral wings protrude. The thoracic wall, relatively thick at this point, thins gradually to the ragged lower margin. In a few specimens the wall thickens, and constricts slightly at about the median point as though to form a second stricture. Pores are small, variable in size and shape, widely and irregularly spaced, sometimes with tiny prongs on their margin which when well-developed subdivide the pores.

Length of cephalis 25–30 μ, of thorax 60–200 μ, of apical horn 25–60 μ, of wings 15–100 μ; diameter of cephalis 40–50 μ, of thorax 50–65 μ. Dimensions based on twenty-eight specimens.

Discussion. In Moreno Gulch (3518) one specimen was found with two horns and no dorsal wing, otherwise very like ?*Lithomelissa heros*. This is believed to be either an aberrant form or an extremely rare second species. As the specimen illustrated by Campbell and Clark and other specimens in the type material all had one horn and a dorsal wing, the description of *L. heros* Campbell and Clark has been emended to include only specimens with one horn.

Localities. CAS loc. 1144, Marca shale (MBP), Moreno Gulch, Cima Hill hole III (at approximately 306·5, 310, 313·5, 320, 327·5, 330, 340, 356·5, 360, and 383 ft.), CAS loc. 39545, and UCMP loc. A2615.

Illustrated specimens. USNM 157935, M39/3 Moreno Gulch (3518) and USNM 157934, J37/0 CAS loc. 1144, both from upper Maestrichtian Moreno formation, Fresno County, California.

? *Lithomelissa hoplites* sp. nov.

Plate 3, figs. 2*a–c*

Description. Shell of two segments: cephalis hemispherical; thorax proximally pyramidal, triangular in transverse section, distally subcylindrical, almost circular in transverse section with wall expanding and constricting irregularly to a narrowed aperture. Cephalis with sturdy, rough, ridged, pointed horn and small upward-directed tube associated with the vertical spine. Ridges from the horn extend on to the cephalis and thorax, appearing as lamellae when well-developed. They are particularly prominent ventrally where they surround the tube and sometimes develop upward-directed spiny projections. The poreless cephalis is covered with tiny papillae. Internally, six collar pores and an axial spine. Sturdy, ridged, dorsal and primary lateral wings extend from the three points of the triangular thorax. Ridges or lamellae of the cephalis extend along the thorax to the wings, and lesser ridges extend from the wings on to the thorax. Thoracic pores are subcircular to elliptical and, except in the distal portion, subdivided by a meshwork. Surface may be smooth or spiny. The constricted distal margin bears a few dull, short, irregular teeth.

Length of cephalis 35–40 μ, of four complete thoraxes 150–272 μ, of wings 25–95 μ. Width of thorax below triangular portion in sagittal plane 80–100 μ, lateral width 70–85 μ. Dimensions based on fifteen specimens.

Discussion. This species is distinguished from ? *L. amazon* as described under that species.

Localities. CAS loc. 1144, Marca shale (MBP), Moreno Gulch, CAS loc. 39545, Stanford loc. PL2875, and Evitt slides AE26 and 27.

Type specimens. Holotype USNM 157936, M39/0; paratype USNM 157937, M37/4, both from upper Maestrichtian Moreno formation, Moreno Gulch (3518), Fresno County, California.

? *Lithomelissa amazon* sp. nov.

Plate 4, figs. 1*a, b*

Description. Shell of two segments: cephalis hemispherical; thorax proximally pyramidal, triangular in transverse section, distally subcylindrical, circular to elliptical in transverse section. Cephalis poreless, covered with numerous tiny papillae, and bearing a sturdy, three-bladed, long, apical horn and a slightly upward-directed tube. Internally, paired ridges extend from the apical spine to the sides of the vertical spine. Four collar pores and a short axial spine are present. Thorax smooth with sturdy, long three-bladed dorsal and primary lateral wings arising from the points of the triangular part of the thorax. Pores are circular to elliptical, widely and irregularly spaced, the proximal ones sometimes with prongs on their margin. The distal edge is thin and ragged, except for one specimen which had a broken velum closing about ten microns above its termination.

Length of cephalis 25–32 μ, of longest thorax 115 μ, of apical horn 35–65 μ, of wings 45–80 μ, of longest axial spine 25 μ; width of thorax below triangular portion 35–60 μ. Dimensions based on sixteen specimens.

Discussion. This species is distinguished from ? *L. hoplites* by its smaller size and smooth surface.

Localities. CAS loc. 1144, Marca shale (MBP), Moreno Gulch, and Cima Hill hole III (at approximately 330 ft.).

Type specimens. Holotype USNM 157938, L37/0 Moreno Gulch (3517); paratype USNM 157939, N48/2 CAS loc. 1144, both from upper Maestrichtian Moreno formation, Fresno County, California.

Genus *Tetraphormis* Haeckel 1881

Remarks. Forms here tentatively assigned to this genus may be closely related especially to the subgenus *Tetraphormis* (*Enneaphormis*), but further investigation of the cephalic structure and geological history of its type species is necessary before the question can be resolved. They are common in the samples studied, but preservation is poor, and only rare specimens have a well-preserved fragile cephalis. Two species have been selected for description—? *T. leia* because it is common though imperfectly preserved, and ? *T. lobocarena* because the cephalic structures are well-preserved, even though the species is very rare.

? *Tetraphormis leia* sp. nov.

Plate 5, fig. 2

Description. Shell of two segments. Cephalis shallow and surface uneven, with minute circular to subcircular scattered pores. It apparently extends on to the thorax for a distance about one-half to two-thirds of the length of the thorax, joining the thorax (commonly at about the fifth to eighth pore-row) with an irregularly lobed distal margin. In some specimens this position is clearly indicated all around the thorax, in others it is only partial, and in still others there is no evidence for its presence. Internally the collar structures, dorsal and two primary lateral spines, extend from the inner margin of the thorax to join a short, barely distinguishable median bar and form three collar pores. Rarely, when these spines are complete, the central area is obscured by spiny structure which consists, at least partially, of two vestigial secondary lateral spines, an apical spine, and two other upward-directed spines, one on each of the primary lateral spines. It is not certain whether an apical horn and tube are present. The flared thorax, approximately discoidal, rises slightly centrally and turns up at the outer rim. Pores are quadrangular (usually) to subcircular, rectangularly arranged in from five to sixteen (usually ten to thirteen) circular (transverse) rows. Pores smaller distally than proximally. Generally there are one or two more rows ventrally than dorsally. On some specimens, longitudinal intervening bars form as many as fifteen prominent ribs, while in others only a few or no distinguishable ribs are present. Distal margin lamellar, narrow and slightly uneven.

Diameter of cephalis 110–247 μ (most 160–230 μ), of collar stricture in sagittal plane 55–75 μ, of thorax 155–450 μ (most 210–350 μ). Dimensions based on thirty specimens.

Discussion. No entirely complete specimen was found and only a few had fragments of the cephalis attached. It is described, because fragments of the thorax are common in three of the four upper Maestrichtian localities studied. There is no definable collar stricture, the base of the cephalis being situated distally from the top of the thorax. It may be that some of the incomplete specimens included (those with no evidence for a broad cephalis on the thorax) belong to another species with a more normal cephalis, its base approximately coinciding with the proximal edge of the thorax.

Localities. CAS loc. 1144, Marca shale (MBP), Moreno Gulch, and Cima Hill hole III (at approximately 356·5 ft.).

Holotype. USNM 157940, L37/3 from upper Maestrichtian Moreno formation, Marca shale (MBP), Fresno County, California.

? Tetraphormis lobocarena sp. nov.

Plate 5, fig. 3; text-fig. 1, fig. 3a, b

Description. Shell of two segments, large lobed cephalis and flared thorax. Cephalis shallow, about twice as broad as high, poreless and with tiny apical horn and tube. Its distal lobed margin extends on to the thorax approximately to the first and second row of pores. Internal structures very distinct—three collar pores formed by the well-developed dorsal and primary lateral spines and median bar, and vestigial vertical and secondary lateral spines are present. The indentations separating cephalic lobes are formed by internal ribs arising at the margin of the cephalis at points collinear with the six main structural spines. These ribs are shortly incorporated into the wall of the cephalis and continue as folds as illustrated by the dotted lines in text-fig. 1, fig. 3*a*. Not all the ribs and folds are constantly present. Besides the apical spine, there are two upright branches, one about midway along each primary lateral spine. Thorax flared, approximately discoidal. Pores subcircular to quadrangular, arranged approximately rectangularly in six to seven (possibly more) circular (transverse) rows. Distal pores smaller and more numerous than median pores. Some of the longitudinal intervening bars in line with the six main structural elements of the collar region are heavy and appear as ribs. The distal margin, where preserved, is as narrow as the intervening bars, fragile, lamellar and slightly uneven.

Height of cephalis 35 μ; diameter of cephalis 75–80 μ, of collar stricture 60–5 μ, of thorax 155–75 μ. Dimensions based on three specimens.

Discussion. Although only three specimens were found, this species is described because of the excellent preservation of the cephalis. The main structural elements of the cephalis are probably similar (though less sturdy) in the many other species thought to belong to this genus in this fauna.

It is distinguished by its smaller cephalis from ? *T. leia*, and by its lobed cephalis from *Sethocephalus clathratus* Koslova (Koslova and Gorbovetz 1966).

Locality. Moreno Gulch.

Holotype. USNM 157941, S32/1 from upper Maestrichtian Moreno formation, Moreno Gulch (3518), Fresno County, California.

Genus *Theocapsomma* Haeckel 1887 emend.

Type species. Theocapsa linnaei Haeckel 1887, p. 1429, pl. 66, fig. 13.

Remarks. The type species of *Theocapsa* (*Theocapsomma*), *T. linnaei* Haeckel, seems to be sufficiently different from that of *Theocapsa* (*Theocapsa*), *T. gratiosa* (Rüst 1885, p. 309, pl. 12, fig. 16), in that the pores of the latter are in regular transverse rows, to justify the separation of these two taxa as genera rather than subgenera.

There are in the upper Maestrichtian of California a number of closely related species with characters which, according to Haeckel's system, would require their placement in separate genera. They are now all included under *Theocapsomma* and the definition of that genus emended to include species with or without small apical horns, and with or without a small constricted aperture. Pores of the thorax and abdomen may or may not be similar, but none are assigned to *Theocapsura* or *Theocapsilla* because this evidently varies intraspecifically in the material here described, and because the type species of these latter have a pronounced neck between cephalis and thorax.

Theocapsomma comys sp. nov.

Plate 4, figs. 2a–c

Description. Shell of three segments: cephalis poreless, small, partly hidden, with four collar pores; thorax conical; abdomen subcylindrical, closed, very variable in shape and size. The degree to which the cephalis is hidden varies, but it is never completely hidden as in *Cryptocapsa asymmetros*, the apical portion generally enclosed in the thoracic wall so that externally there is no collar stricture. Primary lateral and dorsal spines extend into the thoracic cavity. Thorax, circular to slightly elliptical in transverse section, with uniform circular pores (sometimes recessed in circular frames), increasing slightly in size distally, and arranged quincuncially in longitudinal rows, frequently with ridges between. The longitudinal rows of pores and the ridges, when present, continue on to the abdominal segment and the lumbar stricture, while not marked externally by any change in contour, is generally marked by some larger pores formed by the coalescing of pores from the thorax and the abdomen. Abdomen, 3·5–0·3 times as long as thorax, is circular to slightly elliptical in transverse section and in the longer specimens frequently flattened and broken distally. In shorter, complete specimens the abdomen tapers gently to a point which may be asymmetric, or to a smooth rounded or, rarely, a flattened base. In occasional specimens the abdomen is slightly expanded or constricted at some levels, so that it is not always completely symmetrical. In the distal part of the abdomen the pores are larger and less regularly arranged than in the section immediately above.

Length of complete specimens 118–257 μ, of cephalis 15–20 μ, of cephalis and thorax 50–80 μ, of complete abdomen 15–185 μ; diameter of thorax 60–85 μ, of complete abdomen 55–90 μ; of thoracic pores 3–5 μ, of abdominal pores 3–7 μ. Dimensions based on twenty-eight specimens.

Discussion. This species is quite variable in general form, only because of the variability of the abdomen. The thorax, while much more uniform in size and shape, varies in that it tends to be long in those specimens having a short abdominal segment. It differs from other similar species in the vertical alignment of the pores on both thorax and abdomen and/or the lack of an aperture.

Localities. CAS loc. 1144, Marca shale (MBP), Moreno Gulch, Cima Hill hole III (at approximately 310, 313·5, 320, 323·5, 327·5, 330, 340, 356·5, 360, 363·5, and 383 ft.), and ? Evitt slide AE26.

Type specimens. Holotype USNM 157942, Q37/0; paratypes USNM 157943, M39/0, M40/0, all from upper Maestrichtian Moreno formation, CAS loc. 1144, Fresno County, California.

Theocapsomma legumen (Campbell and Clark) emend.

Plate 4, fig. 5

1944*b* *Dictyocephalus* (*Dictyocryphalus*)? *legumen* Campbell and Clark, p. 28, pl. 7, figs. 1, 12, 13, 14.

Description. Shell of three segments: small, partly hidden, poreless cephalis; short, subconical to subcylindrical thorax; and abdomen, very variable in shape and size, with aperture. Primary lateral and dorsal spines extend from cephalis into thoracic cavity and join thoracic wall. Four collar pores present and in some specimens a short delicate axial spine can be distinguished. Pores of the thorax are circular and uniform in size and distribution, rarely recessed in weak angular frames. Externally, lumbar stricture sometimes marked by a slight enlargement of the pores; there is no or only a very slight constriction. Abdomen cylindrical, generally expanded slightly distally, or in the longer specimens flattened and broken distally. In the longer complete specimens the abdomen constricts and curves to form a hook-like termination with a ventrally directed aperture. In shorter specimens the base tends to be more flattened. Pores circular, similar in size and distribution to the thorax proximally, smaller more closely spaced distally. On some specimens a tendency toward transverse alignment in the area where the abdomen begins to constrict. All specimens except those with very short abdominal segments tend to bend toward the ventral side.

EXPLANATION OF PLATE 4

Figs. 1*a, b.* ? *Lithomelissa amazon* sp. nov. *a*, paratype USNM 157939, ventral view. *b*, holotype USNM 157938, left lateral view.

Figs. 2*a–c.* *Theocapsomma comys* sp. nov. *a*, holotype USNM 157942. *b, c*, paratypes USNM 157943; *b*, M39/0; *c*, M40/0.

Fig. 3. *Theocapsomma ancus* sp. nov., holotype USNM 157949, right lateral view.

Fig. 4. *Theocapsomma teren* sp. nov., holotype USNM 157948, right lateral view.

Fig. 5. *Theocapsomma legumen* (Campbell and Clark), USNM 157944, left lateral view.

Fig. 6. *Solenotryma* sp. USNM 157951, incomplete specimen.

Fig. 7. cf. *Solenotryma dacryodes*, USNM 157952.

Fig. 8. *Solenotryma dacryodes* sp. nov., holotype USNM 157950, complete specimen.

Figs. 9*a–c.* *Theocapsomma amphora* Campbell and Clark. *a*, USNM 157947, ventral view. *b*, USNM 157945, right lateral view. *c*, USNM 157946, right lateral view.

Figs. 10*a–d.* *Calyptocoryphe cranaa* sp. nov. *a*, paratype USNM 157955, specimen without small thorax. *b–d*, holotype USNM 157954; *d*, surface detail.

Figs. 11*a, b.* *Hemicryptocapsa conara* sp. nov. *a, b*, holotype USNM 157953; *b*, surface detail.

All figures ×251 except figures 10*d* and 11*b*, ×415.

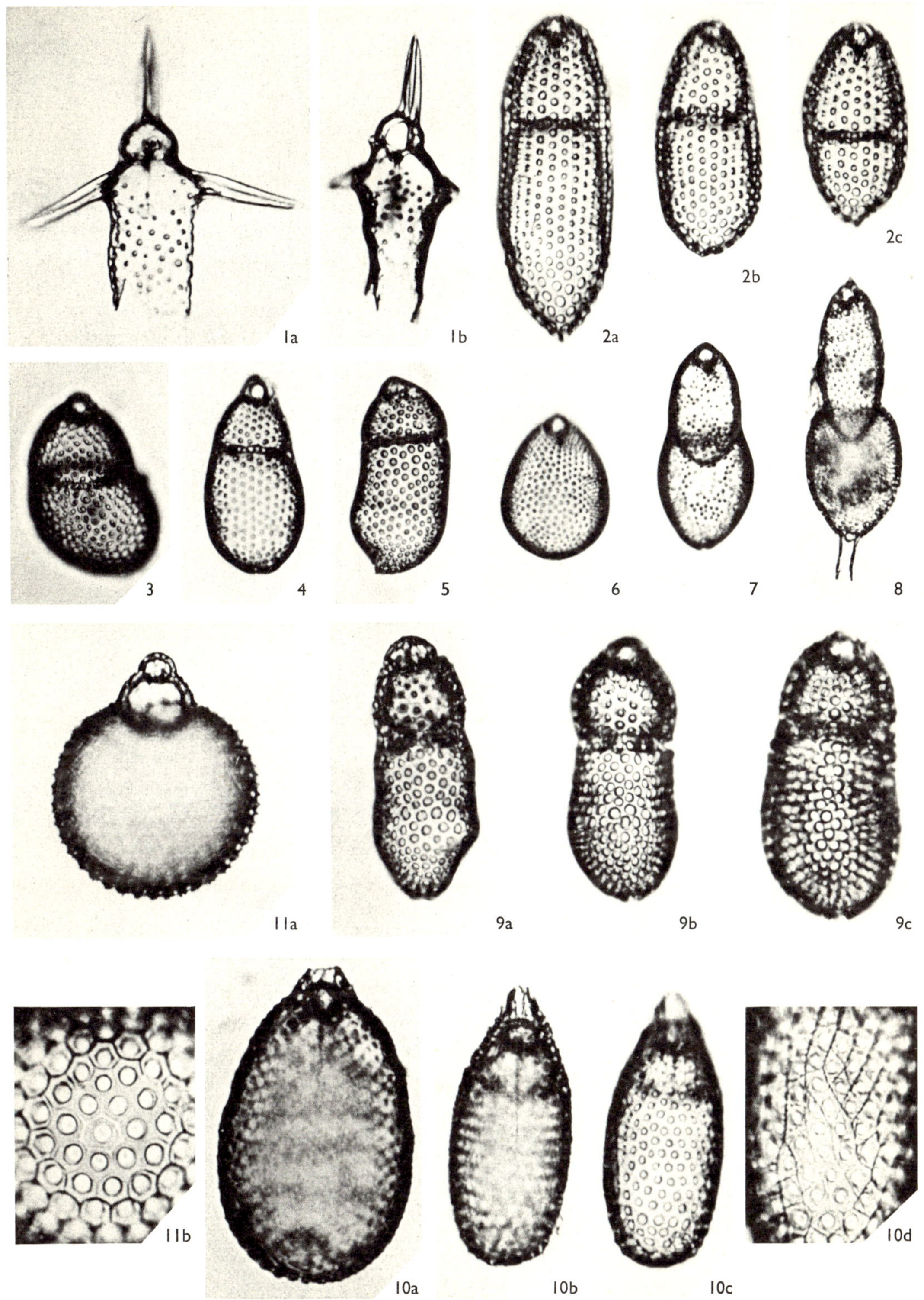

FOREMAN, Upper Maestrichtian Radiolaria of California

Length of complete specimens 100–75 μ, of longest specimen with flattened broken abdomen 230 μ, of cephalis 15–20 μ, of cephalis and thorax together 40–60 μ, of complete abdomen 50–125 μ, of longest broken abdomen 180 μ; diameter of thorax 50–65 μ, of complete abdomen 60–75 μ. Dimensions based on twenty-five specimens.

Discussion. The abdomen of this species varies in shape and size. As in *T. comys* the thorax tends to be large in those specimens with a short abdomen.

Examination of the type material from UCMP loc. A2615 reveals that *Dictyocephalus legumen* does indeed have a small cephalis, partly hidden in the thorax, as pointed out by Deflandre (1953) and as suggested by Campbell and Clark (1944*b*) when they stated, 'the assignment of our species to *Dictyocephalus* is somewhat doubtful since, at least in some specimens, there appears to be an internal, bubble-like structure which actually may be the cephalis of the shell in reduced form withdrawn inside the thorax'. This species is easily distinguished by the characteristic hook-like termination with ventrally directed aperture.

Localities. CAS loc. 1144, Marca shale (MBP), Moreno Gulch, Cima Hill hole III (at approximately 313·5 and 360 ft.), CAS loc. 39545, UCMP loc. A2615, Stanford loc. PL2875, and Evitt slide AE26.

Illustrated specimen. USNM 157944, L38/0 from upper Maestrichtian Moreno formation, Moreno Gulch (Water Canyon), Fresno County, California.

Theocapsomma amphora Campbell and Clark emend.

Plate 4, figs. 9*a–c*

1944*b* *Theocapsa* (*Theocapsomma*) *amphora* Campbell and Clark, p. 35, pl. 7, figs. 30, 31.
1944*b* *Tricolocapsa* (*Tricolocapsium*) *granti* Campbell and Clark, p. 35, pl. 7, figs. 37, 38.

Description. Shell of three segments; cephalis small with four collar pores, thorax conical to subglobose, and abdomen very variable in size and shape with small circular aperture. Cephalis poreless or rarely with a few, small, scattered pores, surface slightly roughened, generally without horns, but occasionally bearing a short stout vertical horn and/or an apical horn which is either small, thorn-like, or short and stout. Cephalis, while not hidden in the thoracic cavity, is partly or rarely completely enclosed in the thick thoracic wall; generally there is a distinct change in external contour from cephalis to thorax. Thorax thick walled with circular pores set in angular, rarely subcircular frames. While there is considerable variation in pore size among specimens, thoracic pores on individuals are uniform in size and arrangement. Abdomen varies considerably in size and shape, from elongate globose to short inverted conical, and tends to be somewhat asymmetric. Most specimens examined have abdominal walls thinner than those of the thorax with pores larger, circular, evenly spaced proximally, and smaller, more closely spaced distally; a few have pores smaller than those of the thorax; most have prongs on their margin which sometimes join to subdivide the pore. Some few specimens have walls as thick as, and pores and frames very like, those of the thorax.

Length (excluding apical horn) 90–240 μ, of cephalis 18–25 μ, of thorax 35–50 μ, of abdomen 45–185 μ; diameter of thorax 52–80 μ, of abdomen 55–115 μ. Dimensions based on forty-five specimens.

Discussion. Although Haeckel considered the similarity or dissimilarity of thoracic and abdominal pores a basis for subgeneric distinction, all gradations between similar and dissimilar pores exist in this species and here, at least, it does not seem practical to make a division on this basis. Besides the dissimilarity of pores Campbell and Clark distinguished *Theocapsa amphora* and *Tricolocapsa granti* by the presence in the former and absence in the latter of a small horn. This is a feature that varies in many of the species here described under *Theocapsomma* and is not considered sufficient reason to separate them from each other. Their apparently thinner more fragile shells are not considered sufficient reason to separate them from the species described above.

Localities. CAS loc. 1144, Marca shale (MBP), Moreno Gulch, Cima Hill hole III (at approximately 310, 313·5, 320, 323·5, 327·5, 330, 340, and 360 ft.), CAS loc. 39545, Stanford loc. PL2875, and Lamont V–18–129.

Illustrated specimens. USNM 157945, M38/0 Moreno Gulch (3518), USNM 157946, N39/4 Moreno Gulch (3518), and USNM 157947, M38/0 CAS loc. 1144, all from upper Maestrichtian Moreno formation, Fresno County, California.

Theocapsomma teren sp. nov.

Plate 4, fig. 4

Description. Shell of three segments; small poreless cephalis, conical thorax, and subcylindrical abdomen with small circular aperture. Cephalis has a thick wall which in many specimens is apically even more thickened to form a broad-based, short, stout apical horn; a few specimens also have a short, stout vertical horn. Collar stricture is not well-defined externally, and internally there are four collar pores. Thorax and abdomen are very similar in character. Walls are of approximately equal thickness, smooth; pores are circular, regular in size and generally irregularly arranged, with sometimes a tendency towards longitudinal alignment. Thoracic pores and those of the proximal part of abdomen frequently enclosed in weak circular or subangular frames. The lumbar stricture is marked by no or only a very slight indentation, so that the cephalis, thorax, and abdomen together are ovate in shape. While some specimens have a straight vertical axis, there is in others a slight tendency for it to be curved with ventral side concave.

 Length (excluding thickened apex of cephalis which varies from 3 to 8 μ) 97–170 μ, of cephalis 15–20 μ, of thorax 30–45 μ, of abdomen 45–113 μ; diameter of thorax 47–60 μ, of abdomen 52–75 μ. Dimensions based on twenty specimens.

Discussion. This species differs from *Tricolocapsa oblonga* Squinabol (1904) in having abdomen shorter, less inflated, and strictures less indented.

Localities. CAS loc. 1144, Marca shale (MBP), Moreno Gulch, CAS loc. 39545, and Stanford loc. PL2875.

Holotype. USNM 157948, N39/0 from upper Maestrichtian Moreno formation, CAS loc. 1144, Fresno County, California.

Theocapsomma ancus sp. nov.

Plate 4, fig. 3

Description. Shell of three segments: small, thick walled, poreless cephalis; conical thorax; and skewed, inversely conical abdomen with small circular aperture. On most specimens the thick rough wall of the cephalis is apically thickened even more to form a short stout horn and some also have a short vertical horn; rarely a protrusion dorsally marks the dorsal wing. In most specimens the collar stricture is not well-defined externally; internally four collar pores are present. Pores of the thorax are uniform, small, circular, generally set in circular or angular frames, evenly distributed quincuncially, tending to show longitudinal alignment. Externally little or no change of contour at the lumbar stricture. Pores of the abdomen are small, circular, uniform in size, generally irregularly distributed throughout or tending to become slightly larger and more closely spaced distally, frequently set in discrete circular or angular frames which rarely coalesce to enclose more than one pore. The eccentric abdomen is markedly skewed toward the ventral side, except in a few specimens where the distortion is very slight.

Length (not including thickened apex of cephalis which varies from 3 to 8 μ) 100–35 μ, of cephalis 15–20 μ, of thorax 35–45 μ, of abdomen 45–95 μ; diameter of thorax 55–75 μ, of abdomen 55–90 μ. Dimensions based on twenty specimens.

Discussion. This species is distinguished from *T. teren* by its skewed abdomen and rougher surface.

Localities. CAS loc. 1144, Marca shale (MBP), Moreno Gulch, and Cima Hill hole III (at approximately 310 ft.).

Holotype. USNM 157949, N37/4 from upper Maestrichtian Moreno formation, CAS loc. 1144, Fresno County, California.

Genus Solenotryma gen. nov.

Type species. *S. dacryodes* sp. nov.

Diagnosis. Eradiate tricyrtid forms with small, simple cephalis, relatively large thorax with constricted aperture, and abdomen with constricted (sometimes tubular) aperture.

This genus differs from other tricyrtids with constricted terminal aperture (*Theocapsomma* Haeckel, *Theocampe* Haeckel, and *Eusyringium* Haeckel) in that the aperture at the base of the thorax is also strongly constricted, and the abdomen is more an appendage than a fundamental part of the skeleton.

Although the species described below might appear to conform to the generic definition of *Theocorys* as written by Haeckel (1881) it is not assigned to that genus because it is evidently not related to its type species. The type species of *Theocorys* (*Theocorys*) is, by monotypy, *T. morchellula* Rüst (1885), and it is believed that *S. dacryodes* should be generically separated from this because the abdomen in Rüst's form appears to be a more important and stable element of the skeleton.

Etymology. From the Greek *solen*, tube and *tryma*, pore (neuter).

Solenotryma dacryodes sp. nov.

Plate 4, fig. 8

Description. Shell of three segments; cephalis and thorax together ovate to ellipsoidal, abdomen very variable in shape and wall thickness, frequently inverted conical, with a slender tube distally. The small poreless cephalis is partly hidden in the thorax. An apical horn is variably developed; it may be entirely lacking, or represented by a group of rough nodes, tiny thorn, or slender spine. Internally, there are four collar pores and an axial spine; the primary lateral and sometimes the dorsal spines extend into the thoracic cavity to join the thoracic wall. Thorax generally smooth with small, circular to elliptical, irregularly arranged pores varying slightly in size, closely or widely spaced; basally with a small, circular, smooth opening, the distal part of the thorax hidden in the abdominal cavity. Abdomen with circular to subangular, closely spaced, regular pores; basally with a slender tube which has a few irregular pores and tiny lamellar teeth marginally.

Length of cephalis 13–15 μ, of thorax and cephalis together 70–105 μ, of abdomen 27–110 μ, of tube 10–25 μ, of well-developed apical horn 18–25 μ; diameter of thorax 45–75 μ, of abdomen 32–75 μ, of tube 8–12 μ. Dimensions based on eight complete specimens.

Discussion. This distinctive species varies considerably.

In the upper Maestrichtian, only one specimen (from Moreno Gulch 3517) has been observed with three segments and no tube (Pl. 4, fig. 7); it is possible that when well-developed or complete a fourth segment with tube may be present. From the lower Maestrichtian–? upper Campanian, Campbell and Clark (1944*b*) illustrated two specimens, plate 7, figs. 10 and 11, *Nodosaria* sp. and *Nodosaria cf. velascoensis* Cushman, the former with three, the latter with five segments, which they considered as silicified Foraminifera although the radiolarian cephalis can be distinguished in both drawings. Another specimen seen in topotypic material (UCMP loc. A2615) had four segments and a fragment of a tube. A single specimen, possibly a member of this species, has been observed from Lamont V–18–129. In all of the above mentioned specimens the last-preserved segment had pores similar in character to those of the thorax and lacked a clearly defined tube, though an additional segment with tube might originally have been present. These specimens are not now included in the above described species because it is suspected that they may represent an earlier species with a greater number of segments and are referred to herein as cf. *Solenotryma dacryodes*.

In the upper Maestrichtian samples CAS loc. 1144, Marca shale (MBP), Moreno Gulch and Cima Hill hole III (at approximately 327·5, 360, and 383 ft.) specimens with the abdomen lacking and the thorax either entirely smooth, or with a ring of tiny thorns distally to indicate its former presence are common (length of thorax and cephalis 60–125 μ, diameter of thorax 45–75 μ, of aperture 6–8 μ; dimensions based on thirty specimens). They are here referred to as *Solenotryma* sp. (Pl. 4, fig. 6) and are believed to belong to *S. dacryodes*, but cannot be confidently identified because of the possibility that they could also be incomplete specimens of the specimens discussed above and referred to as cf. *Solenotryma dacryodes*. The very rare incomplete specimens like those described above present in the lower Maestrichtian–? upper Campanian CAS loc. 39545 probably are incomplete specimens of cf. *Solenotryma dacryodes*.

Sethocapsa microacantha Squinabol (1903) and *Halicapsa crebripora* Squinabol (1904) (which probably has a small cephalis), superficially resemble *Solenotryma* sp. but differ in lacking an aperture.

Localities. CAS loc. 1144, Marca shale (MBP), and Moreno Gulch.

Holotype. USNM 157950, W53/3 from upper Maestrichtian Moreno formation, Marca shale (MBP), Fresno County, California.

Illustrated specimen. Solenotryma sp. USNM 157951, O37/0 from upper Maestrichtian Moreno formation, CAS loc. 1144, Fresno County, California.

Illustrated specimen. Cf. *Solenotryma dacryodes* USNM 157952, M37/2 from upper Maestrichtian Moreno formation, Moreno Gulch (3517), Fresno County, California.

Genus *Hemicryptocapsa* Tan Sin Hok 1927 emend.

Type species. H. capita Tan Sin Hok 1927, p. 50, pl. 9, fig. 67.

Diagnosis. Tricyrtid forms with large inflated spherical or subspherical abdomen, the latter with a distinct relatively large pore or tube near its junction with the thorax. Cephalis free, with or without a small apical horn, and thorax completely or only partly hidden in abdomen; abdomen with or without aperture.

This genus is distinguished from all other superficially similar genera by its partly or completely hidden thorax and the large pore or tube near the upper margin of the abdomen.

Hemicryptocapsa conara sp. nov.

Plate 4, figs. 11*a, b*

Description. Shell of three segments: small, poreless cephalis; small, poreless, partly hidden thorax; and large, spherical to subspherical abdomen without aperture. Cephalis only slightly roughened or with tiny spines or ridges; either without horns or with small apical horn and rarely tiny vertical horn; collar stricture well-defined; internally four collar pores and an axial spine, which when well-developed extends slightly beyond the thorax. Thorax rough with irregular ridges and nodes, sometimes with a few short, sharp spines; lower margin has short smooth collar with straight sides and a ring of slender, outwardly directed, curved spines immediately above. These spines possibly join inner wall of abdomen. Abdomen with circular pores of uniform or somewhat variable size, set in angular frames. A single large elliptical pore or tube is situated on the ventral side immediately below the junction of the thorax and abdomen. Most specimens have tiny spines at the angles of the frames, and some a few scattered, sharp, more prominent spines.

Length of cephalis 15–22 μ, of thorax 35–45 μ, of abdomen 55–140 μ; diameter of thorax 42–55 μ, of abdomen 60–150 μ, of pores 5–10 μ, of frames 8–15 μ. Dimensions based on twenty specimens.

Discussion. H. conara may be conspecific with *Tricolocapsa abdominalis* Rüst in Iacob and Nicorici (1957, p. 13, p. 19, fig. 16) but the figure is in oblique longitudinal section and it is therefore not certain to what extent the thorax is hidden in the abdomen. It is distinguished from *Tricolocapsa pilula* Hinde (1900) and *T. pilula* Hinde (1908) by its thorax which is longer and not so deeply hidden. It differs from *T. sphaeroides* Neviani (1900) in the latter having cephalis and thorax together spherical.

Localities. CAS loc. 1144, Marca shale (MBP), Moreno Gulch, and Cima Hill hole III (at approximately 320 and 340 ft.).

Holotype. USNM 157953, M35/0 from upper Maestrichtian Moreno formation, Moreno Gulch (3518), Fresno County, California.

Genus *Calyptocoryphe* gen. nov.

Type species. C. cranaa sp. nov.

Diagnosis. Tricyrtid forms occasionally lacking a thorax, with thorax (when present) partly or completely hidden in abdomen. Cephalis and upper part of thorax enclosed in an apically open sheath which arises from the thoracic or abdominal wall.

This genus is distinguished from *Hemicryptocapsa* in that there is no differentiated larger pore or tube proximally on the last segment, and the latter is more irregular in outline.

Etymology. The name is derived from the Greek *kalyptos*, covered and *koryphe*, head (feminine).

Calyptocoryphe cranaa sp. nov.

Plate 4, figs. 10*a–d*

Description. Shell of three (or rarely two) segments; small poreless cephalis usually partly hidden in a small hemispherical thorax, which, in turn, is partly or almost wholly hidden in a large ellipsoidal to globular abdomen without an aperture. Two specimens have been observed alike in every respect except that they lack the thorax, so that the second segment is apparently homologous with the abdomen described below (Pl. 4, fig. 10*a*). A hollow, rough, irregularly ridged, generally poreless sheath encloses the proximal portion of the thorax, sometimes appears to merge with the thoracic wall and extends upward to surround the cephalis asymmetrically, so that the apical horn is incorporated in its wall. This sheath narrows and remains open apically and eccentrically. Internally, the cephalis has four collar pores, and a long axial spine extends into the abdomen. A short vertical horn is sometimes present on the cephalis, and rarely secondary lateral wings at the collar stricture. On the thorax, dorsal and primary lateral wings are occasionally present, and rarely other scattered spines. Pores are sometimes present at the base of the sheath. The hidden portion of the thorax has a thin wall with uniform circular pores. Pores of the abdomen vary with its shape. Ellipsoidal specimens have circular to subcircular pores, quincuncially arranged with rows separated by vertical or diagonal ridges and swirled frames around each pore, the pores becoming markedly larger distally. Specimens with a more globular abdomen have no, or less prominent, ridges with circular pores quincuncially arranged set in angular frames, very little swirling, less tendency for the pores to be arranged in rows, and on some specimens a few larger basal pores.

Length of three segments 133–212 μ, of cephalis 18–25 μ, of thorax 30–55 μ, of abdomen 100–75 μ; diameter of thorax 45–65 μ, of abdomen 65–155 μ. Dimensions based on twenty specimens.

Discussion. The variation in the presence of, and in the position of the thorax in relation to the abdomen and the range in shape and size of the latter cause considerable variation in appearance.

Localities. CAS loc. 1144, Marca shale (MBP), and Moreno Gulch.

Type specimens. Holotype USNM 157954, L40/0 CAS loc. 1144; paratype USNM 157955, N38/1 Moreno Gulch (3518), both from upper Maestrichtian Moreno formation, Fresno County, California.

Genus *Myllocercion* gen. nov.

Type species. *M. acineton* sp. nov.

Diagnosis. Tricyrtid forms lacking well-developed apical horn, with basally constricted abdomen and ribs in or inside of abdominal wall which may or may not be prolonged into three lamellar feet.

Myllocercion differs from the type species of *Pleuropodium* Haeckel (1881) in the former lacking a well-developed apical horn and having a constricted abdomen; feet when present are lamellar. It differs from the type species of *Podocyrtis* Ehrenberg (1847*a*) and *Thyrsocyrtis* Ehrenberg (1847*b*) in lacking a well-developed apical horn and having three ribs in or inside of the abdominal wall.

Etymology. The name is derived from the Greek *myllos*, fish and *kerkos*, tail (neuter).

Myllocercion acineton sp. nov.

Plate 5, figs. 11*a, b*

Description. Shell of three segments; cephalis small poreless, thorax conical to globose, and abdomen inverted conical to globose, triangular in transverse section, with a small circular aperture. The cephalis of some specimens shows a very slight thickening at the apex and one specimen had a short, stout, vertical horn; surface with numerous small depressions. The thoracic wall frequently encloses the lower part of the cephalis and as a result the collar stricture with four collar pores is not always clearly defined externally. The thorax has wavy vertical ridges generally joined by short transverse ridges forming rectangular to rounded frames quincuncially arranged. Single, small, circular pores are situated in the frames at and near the collar and lumbar strictures. The abdomen has a thin wall with numerous tiny circular to ellipsoidal pores tending to form longitudinal rows. Three, generally broad, lamellar feet, or ribs (?) with irregular wavy ridges, resembling somewhat a fish-tail, are incorporated in the wall of the abdomen. These constrict rapidly, indenting the abdomen distally to form the triangular transverse section mentioned above. This abdomen is very fragile and rarely preserved whole. Most specimens consist of cephalis, thorax, and stubs of three feet.

Length of three segments 105–33 μ, of cephalis 20–2 μ, of cephalis and thorax 70–85 μ, of abdomen 25–70 μ; diameter of thorax 55–72 μ, of abdomen 35–75 μ, of aperture 5–8 μ. Dimensions based on twelve specimens, except length of three segments and abdominal measurements which are based on seven.

Discussion. It is virtually impossible to distinguish this species confidently from *M. rhodanon* when the distinctive abdomen and feet are missing. Incomplete specimens which may belong to either of the two species described are common in the upper Maestrichtian of Fresno County, California, and present also in CAS loc. 39545, Stanford loc. PL2875, and Lamont V–18–129.

Localities. CAS loc. 1144, Marca shale (MBP), Moreno Gulch, Cima Hill hole III (at approximately 320, 327·5, 340, and 360 ft.), UCMP loc. A2615, CAS loc. 39545, and Stanford loc. PL2875.

Type specimens. Holotype USNM 157956, M38/0; paratype USNM 157957, M40/0, both from upper Maestrichtian Moreno formation, CAS loc. 1144, Fresno County, California.

Myllocercion rhodanon sp. nov.

Plate 5, fig. 12

Description. Shell of three segments; cephalis and thorax exactly as in *M. acineton*, abdomen inflated to cylindrical. Wavy ridges continue from the thorax on to the abdomen, sometimes with transverse ridges between them forming angular to rounded frames quincuncially arranged in longitudinal rows as on the thorax. These ridges and frames become less regular distally and in some specimens are weak or missing entirely. Small circular pores are situated in the frames below the lumbar stricture, distally above the aperture, or in all the frames. Inside the abdomen, three slender ribs extend from the lumbar stricture part way or completely to the aperture; it is not certain whether they are free or attached to the inner wall. The constricted aperture is large and has a smooth circular rim. Three, short, equidistant, broad-based, triangular, lamellar feet, sometimes with wavy lines characteristic of the feet of *M. acineton*, generally in line or connected with the three ribs mentioned above, extend downwards from the distal edge of the abdominal wall. Single smaller secondary feet are sometimes present between the three mentioned above.

EXPLANATION OF PLATE 5

Figs. 1*a*, *b*. ? *Clathrocyclas hyronia* sp. nov., holotype USNM 157974, right lateral view.
Fig. 2. ? *Tetraphormis leia* sp. nov., holotype USNM 157940, apical view, dorsal spine at right.
Fig. 3. ? *Tetraphormis lobocarena* sp. nov., holotype USNM 157941, basal view, dorsal spine at right.
Fig. 4. ? *Clathrocyclas diceros* sp. nov., holotype USNM 157972, right lateral view.
Figs. 5*a*, *b*. ? *Clathrocyclas lepta* sp. nov. *a*, *b*, holotype USNM 157973. *a*, right lateral-ventral view, showing stub of secondary lateral wing at collar stricture; *b*, left lateral view.
Figs. 6*a–d*. ? *Ectonocorys hemicarena* sp. nov. *a*, *b*, holotype USNM 157965, left lateral view. *c*, paratype USNM 157967, ventral view. *d*, paratype USNM 157966, apical view.
Figs. 7*a*, *b*. *Ectonocorys lampra* sp. nov., holotype USNM 157960, left lateral view.
Figs. 8*a–c*. *Ectonocorys scolia* sp. nov. *a*, *b*, holotype USNM 157961, left lateral view. *c*, paratype USNM 157962, apical view.
Figs. 9*a*, *b*. *Ectonocorys deltota* sp. nov. *a*, paratype USNM 157964, apical view. *b*, holotype USNM 157963, left lateral view.
Figs. 10*a*, *b*. *Cornutovum levigatum* sp. nov., holotype USNM 157959, left lateral view.
Figs. 11*a*, *b*. *Myllocercion acineton* sp. nov. *a*, paratype USNM 157957. *b*, holotype USNM 157956.
Fig. 12. *Myllocercion rhodanon* sp. nov., holotype USNM 157958.
Figures 1–5, ×166; 6–12, ×251.

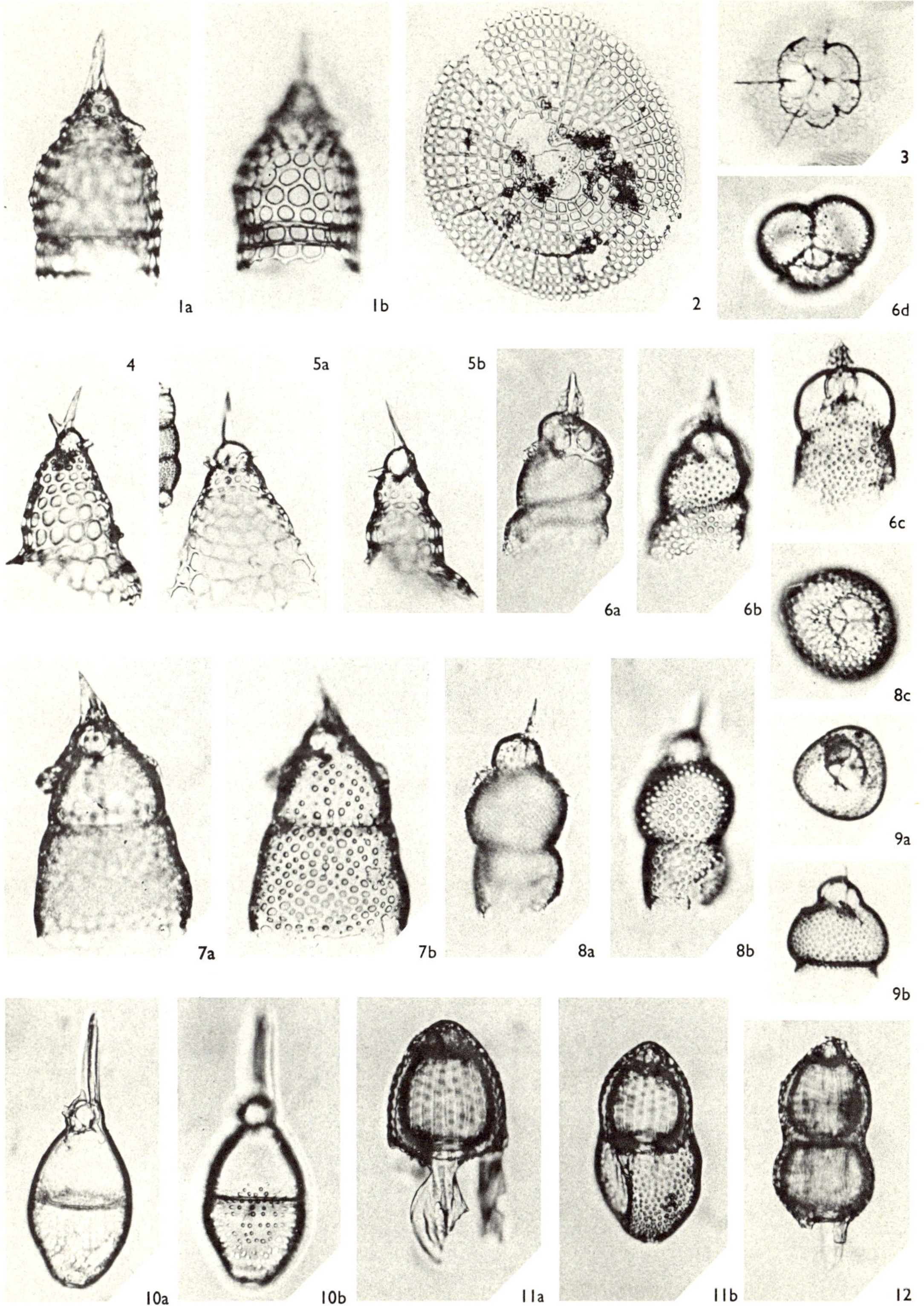

FOREMAN, Upper Maestrichtian Radiolaria of California

Length (exclusive of basal feet) 105–45 μ, of cephalis 18–22 μ, of cephalis and thorax 60–78 μ, of abdomen 40–65 μ, of basal feet 6–35 μ; diameter of thorax 50–75 μ, of abdomen 58–80 μ, of aperture 20–5 μ. Based on eight specimens, all with complete abdomens.

Discussion. This species is distinguished from *M. acineton* by the differences in the character of the abdominal segment and the absence of lamellar feet beyond the abdomen in the latter. Complete specimens are very rare.

Localities. CAS loc. 1144, Marca shale (MBP), and Moreno Gulch.

Holotype. USNM 157958, N38/0 from upper Maestrichtian Moreno formation, CAS loc. 1144, Fresno County, California.

Genus *Cornutovum* gen. nov.

Type species. C. levigatum sp. nov.

Diagnosis. Tricyrtid forms with well-developed apical horn and cephalic tube, abdominal pores tending to be transversely aligned, and aperture constricted or closed.

This genus is most obviously distinguished from *Tricolocapsa* Haeckel (1887) by its large cephalic horn.

Etymology. The name is derived from the Latin *cornutus*, horned and *ovum*, egg (neuter).

Cornutovum levigatum sp. nov.

Plate 5, figs. 10*a, b*

Description. Shell smooth, of three segments: cephalis small, poreless, bearing a sturdy, three-bladed apical horn, and an upward-directed tube in line with the vertical spine; thorax conical; and abdomen inversely conical with small circular aperture. Internally, paired ridges extend from the apical spine to the sides of the vertical spine on most specimens. Four collar pores and a short, delicate axial spine are present. Collar stricture well defined. Thorax poreless proximally, occasionally with small round depressions; distally with small, circular to subcircular, unevenly scattered pores. One specimen had a short, lamellar, broad-based spine extending outward from the thorax directly below the vertical spine. No external lumbar stricture; thorax and abdomen together ovate or ellipsoidal. Pores of the abdomen circular to elliptical, very unevenly distributed, with a slight tendency towards transverse alignment.

Length of three segments (exclusive of apical horn) 90–130 μ, of cephalis 15–20 μ, of thorax 25–45 μ, of abdomen 40–70 μ, of apical horn 25–85 μ; diameter of thorax 45–75 μ, of abdomen 50–85 μ, of apical horn 5–12 μ, of thoracic pores $\frac{1}{2}$–3 μ, of abdominal pores $\frac{1}{2}$–5 μ, of aperture 10–15 μ. Dimensions based on twenty specimens.

Discussion. This very distinctive species varies little and resembles no other species in this fauna. Specimens from Marca shale (MBP) and CAS loc. 1144 have smaller pores than those from Moreno Gulch.

Localities. CAS loc. 1144, Marca shale (MBP), Moreno Gulch, and Cima Hill hole III (at approximately 340 ft.).

Holotype. USNM 157959, M38/4 from upper Maestrichtian Moreno formation, Marca shale (MBP), Fresno County, California.

Genus *Ectonocorys* gen. nov.

Type species. E. lampra sp. nov.

Diagnosis. Tricyrtid forms, not flared distally, with an apical horn and distinctive cephalic structure characterized by paired ridges extending from the apical spine along the cephalic wall to the primary lateral spines.

This genus is distinguished from *Theocyrtis* Haeckel by the cephalic structure in that it appears probable that the cephalic structure of *Eucyrtidium barbadense* Ehrenberg (1873) (the type species of *Theocyrtis*) is similar to that described by Riedel (1957) and Nigrini (1967), for *Anthocyrtidium cineraria, Calocyclas virginis, Theocorythium trachelium dianae,* etc. While most of the cyrtoids described in this paper have paired cephalic ridges extending from the apical spine to the sides of the vertical spine (text-fig. 1, figs. 1*a,* 7, 8*a,* 9*a,* and 10*a*) the following three species and the one doubtfully assigned lack the latter and have paired ridges extending from the apical spine to the primary lateral spines (text-fig. 1, figs. 5 and 6).

Etymology. The name is derived from the Greek *ektonos,* out of tune and *korys,* cap (feminine).

Ectonocorys lampra sp. nov.

Plate 5, figs. 7*a, b*

Description. Shell of three or rarely four segments; cephalis and thorax together conical, abdomen subcylindrical. Cephalis with some scattered pores, and a short apical horn with a broad, irregularly ridged base, tapering to a smooth point; occasionally the basal ridges extend upward throughout the length of the horn. Collar stricture not well defined externally. Internally, there are six collar pores, and weak paired ridges extend from the apical spine along the cephalic wall to the primary lateral spines. Thorax elliptical to subcircular in transverse section, and when viewed apically the major axis of the ellipse is frequently skewed in relation to the sagittal plane of the cephalis; its widest dimension is, in most cases, associated with the direction of the sagittal plane. A large rough tube extends downward from the vertical spine for approximately half the length of the thorax. Three small, sharp wings extend from the dorsal and primary lateral spines; these wings sometimes connected to the thorax by bars. Pores subcircular, variable in size, and widely scattered on some specimens; more uniform in size, and more closely and uniformly spaced in others. Lumbar stricture not well defined externally, and generally the width of the abdomen is equal to or greater than that of the thorax. Abdominal pores subcircular, irregularly spaced, slightly larger than those of the thorax. Distal margin always ragged, except for one specimen with a fragment of a short fourth segment that had a margin of irregular short teeth.

Length of cephalis 20–5 μ, of thorax 40–55 μ (most 40–50 μ); width of short axis of thorax 65–80 μ, of abdomen 60–90 μ, of long axis of thorax 70–90 μ, of abdomen 70–115 μ. Dimensions based on seventeen specimens.

Discussion. This species is distinguished from *E. scolia* by its larger size and the ridged character of the apical horn.

Localities. CAS loc. 1144, Marca shale (MBP), Moreno Gulch, and CAS loc. 39545.

Holotype. USNM 157960, O39/3 from upper Maestrichtian Moreno formation, Moreno Gulch (3518), Fresno County, California.

Ectonocorys scolia sp. nov.

Plate 5, figs. 8*a*–*c*

Description. Shell of three segments; cephalis and thorax together campanulate, abdomen subcylindrical. Cephalis poreless or with few scattered pores, bearing a cylindrical apical horn approximately as long as the cephalis; in some specimens horn surrounded by spongy meshwork. Collar stricture well-defined externally. Internal collar structures weak and frequently broken; when complete there are six collar pores. Paired ridges extend from the apical spine to the primary lateral spines along the inner surface of the cephalis. Thorax elliptical to subcircular in transverse section, and when viewed apically appears skewed in relation to sagittal plane of cephalis. A small, rough tube extends downwards from the vertical spine directly below the collar stricture, and three small thorns (wings) extend from the dorsal and primary lateral spines. Pores are subcircular, fairly uniform in size, closely spaced, and irregularly or quincuncially arranged. Tiny spines sometimes present over all. Lumbar stricture well-defined. Abdomen fragile, short, subcylindrical with irregular contour and distal edge ragged. It is, in most cases, as wide as or narrower than the thorax, with pores subcircular, more variable in size than on thorax, tending to be transversely or irregularly arranged.

Length of cephalis 20–5 μ, of thorax 35–60 μ (most 45–50 μ); width of short axis of thorax 60–80 μ, of abdomen 50–75 μ (most 55–65 μ), of long axis of thorax 70–90 μ, of abdomen 60–100 μ (most 70–85 μ). Dimensions based on fifteen specimens.

Discussion. This species is distinguished from *E. deltota* by its size, and the shape and skewness of the thorax.

Localities. CAS loc. 1144, Marca shale (MBP), Moreno Gulch, CAS loc. 39545, and UCMP loc. A2615.

Type specimens. Holotype USNM 157961, N38/2; paratype USNM 157962, N38/4, both from upper Maestrichtian Moreno formation, Moreno Gulch (3517), Fresno County, California.

Ectonocorys deltota sp. nov.

Plate 5, figs. 9*a*, *b*; text-fig. 1, fig. 5

Description. Shell small of three or rarely four segments; cephalis and thorax together campanulate, abdomen inflated, cylindrical, of one or possibly two segments. Cephalis poreless or with a very few small pores, bears an apical horn, generally broken, but when complete, cylindrical, slightly longer than the cephalis, and surrounded by spongy meshwork. A small, delicate tube extends from the vertical spine at the collar stricture a short way down on the thorax. Internally, well-developed ridges extend along the cephalic wall from the apical spine to the primary lateral spines. Six collar pores are present and the dorsal spine has tiny thorns on its lower margin. Externally, the collar stricture is well defined ventrally. Thorax triangular in transverse section with small thorns (wings) extending from the dorsal and primary lateral spines perpendicular to the faces of the triangle. Most specimens have furrows extending downward from the collar stricture in line with these spines and frequently the thoracic wall bulges between them to obscure the collar stricture dorsally. Pores subcircular, fairly uniform in size on individual specimens but vary somewhat among specimens. Most have very small (1–2 μ) pores, widely scattered. Specimens with larger pores tend to have them quincuncially arranged and more closely spaced. Thorax markedly constricted at the lumbar stricture, which tends to slope upward dorsally. Fragile abdomen rarely well-preserved with pores slightly larger and more variable in size than those on thorax and tending toward transverse alignment. Distal edge always ragged, an occasional specimen showing a stricture and fragment of a fourth segment.

Length of cephalis 20 μ, of thorax 35–40 μ (most 35 μ); width in sagittal plane of thorax 55–65 μ, of abdomen 50–65 μ, in lateral plane of thorax 55–70 μ, of abdomen 50–60 μ. Dimensions based on twenty-five specimens.

Discussion. This species is distinguished from *E. scolia* as discussed under that species.

Localities. CAS loc. 1144, Marca shale (MBP), Moreno Gulch, and CAS loc. 39545.

Type specimens. Holotype USNM 157963, M38/2; paratype USNM 157964, N38/4, both from upper Maestrichtian Moreno formation, Moreno Gulch (3517), Fresno County, California.

?*Ectonocorys hemicarena* sp. nov.

Plate 5, figs. 6a–d; text-fig. 1, fig. 6

Description. Shell of three segments; small incomplete cephalis, inflated thorax, and short fragile abdomen with ragged lower margin. Relationship of cephalis to thorax unusual in that the defining collar stricture is developed only between the primary lateral and vertical spines which form only the two cervical collar pores. Secondary lateral spines are short thorns which do not reach the enclosing wall. Apical horn cylindrical, about equal to cephalis in length, with irregular spongy meshwork at its base and sometimes over all. Internally, apical spine sturdy with three branches apically. A short, almost horizontal one joins the dorsal wall and longer, more prominent paired branches or ridges extend along the inner edge of the cephalis to join the primary lateral spines. The cephalic wall is externally furrowed along the course of these latter two apical spine ridges, and it is the area between these two furrows and the abbreviated collar stricture which is considered as the cephalis. The area extending dorsally from the furrows and downward from the collar stricture is considered as the thorax. Besides being outlined by the furrows described above, the reduced cephalis is also differentiated by having no pores or much smaller and more widely spaced pores than the thorax.

A delicate axial spine is present. The sturdy dorsal spine has short thorns on its lower side; as it approaches the thoracic wall it expands vertically and joins the wall with a row of small, vertical, lamellar bars. Externally, the thorax has a deep furrow here. Thorax about one-third narrower in the sagittal plane than in the lateral plane, the dorsal furrow emphasizing this narrowing and giving the thorax a two-lobed appearance. Pores are small, circular to subcircular, variable in size, generally widely scattered, and distinctly larger on that portion of the thorax between the primary lateral spines than dorsally. Tiny thorns (wings) extend from the primary lateral spines and sometimes the dorsal spine, and a small, rough, sometimes inconspicuous, downwardly directed tube protrudes from the vertical spine. Lumbar stricture oblique, low ventrally and high dorsally. Abdominal pores similar to those on the ventral side of the thorax.

Length of cephalis 20–5 μ, of thorax below vertical spine 35–45 μ, of apical horn 20–35 μ; diameter of cephalis 18–27 μ, of thorax along sagittal plane 50–5 μ, along lateral plane 65–80 μ, of thoracic pores 1–5 μ, of abdominal pores 1–5 μ. Dimensions based on ten specimens.

Discussion. This species is distinguished from all other described species in that the collar stricture is not complete, the cephalis reduced, and the thorax extends to the base of the apical horn dorsally. A very rare, similar, undescribed species from CAS loc. 39545 differs in having cephalic pores, larger thoracic pores, and a less well-developed dorsal spine.

Although the reduced cephalis of this species gives it a markedly different appearance from the other species included in *Ectonocorys* its actual close relationship, particularly to *E. deltota*, is shown by the following features: (1) internal apical-primary lateral ridges, (2) triangular shape of thorax with wings emerging from faces of triangle, not corners, (3) loss of collar stricture definition dorsally, (4) furrows on thorax in line with dorsal spine and primary lateral spines, (5) tiny thorns on lower margin of internal dorsal spine, (6) shape and size of pores.

?*E. hemicarena* is doubtfully assigned to this genus because although its cephalic structure is similar in that it has paired ridges extending from the apical spine to the primary lateral spines, its cephalis is reduced and cardinal and jugular pores are lacking.

Localities. CAS loc. 1144, Marca shale (MBP), Moreno Gulch, and Cima Hill hole III (at approximately 330 ft.).

Type specimens. Holotype USNM 157965, M39/1 Marca shale (MBP); paratypes USNM 157966, N39/3 Moreno Gulch (3517), and USNM 157967, N39/0 CAS loc. 1144, all from upper Maestrichtian Moreno formation in Fresno County, California.

Genus *Sciadiocapsa* Squinabol 1904 emend.

Remarks. It appears probable that the species described below are sufficiently closely related to *Sciadiocapsa euganea* (the type species) to require their placement in the same genus. If this is true, the genus is certainly not monocyrtid but rather di- or tricyrtid. It is difficult to define the limits of the thorax and to determine whether or not an abdomen is present. Because there is no internal constriction, no differentiation of the pores, and in some species very little change in contour the flared portion could very well be considered as part of the thorax. Alternatively the velum could be considered part of the thorax homologous to the hidden part of the thorax of *Hemicryptocapsa conara*. No decision is here made regarding this question—it is only for the sake of convenience and brevity that, in the species description below, the flared portion of the shell is described as an 'abdomen'.

They are doubtfully assigned because of uncertainty regarding the internal cephalic structure of *S. euganea*. ?*S. ptesimolecis* has internal paired ridges along cephalic wall from apical to primary lateral spines, and ?*S. causia* and ?*S. petasus* have internal paired ridges from apical to vertical spine. All three have a tube extending from the vertical spine.

?*Sciadiocapsa ptesimolecis* sp. nov.

Plate 7, fig. 3; text-fig. 1, figs. 2*a*, *b*

Description. Fragile shell of three segments and a basal velum which is constricted, rarely completely closed; cephalis and thorax together short conical, and abdomen flared. Cephalis, with surface depressions and a few pores, with tiny apical spine and a small, horizontal, fragile tube. When, as is frequently the case, the tube is broken off a larger than average pore divided by a spine can be seen on the thorax immediately below the collar stricture. Internally, weak paired ridges extend along the cephalic wall from the apical spine to the primary lateral spines; and there are six, frequently broken, collar pores. Little or no change of contour between cephalis and thorax except ventrally. Thorax circular to elliptical in transverse section, with circular to subquadrangular pores arranged in three to four transverse rows and eleven to fifteen longitudinal rows (counted on the proximal thorax). Abdominal pores similar to those of thorax. Abdomen subcircular with greatest width (length) approximately between the right primary lateral and ventral spines, and margin, when complete, rough. Velum originates where abdomen flares (lumbar stricture?); it may be complete, forming an asymmetrical bowl-like segment without aperture, or incomplete. When incomplete the velum is developed approximately between the left primary lateral spine dorsally to the right secondary lateral spine. In the latter case the wall shortens as it approaches the two end points, so that there is no wall or only a few teeth between the right primary lateral and the vertical spine. Surface smooth with small, irregular pores tending to be arranged transversely. Velum, when viewed in profile, is almost collinear with, and appears as a continuation of, the thorax.

Length of cephalis 20–5 μ, of thorax 10–30 μ, of velum 10–40 μ; diameter of thorax 55–70 μ, of greatest width of abdomen 95–118 μ. Dimensions based on fifteen specimens.

Discussion. This species is considerably smaller than *S. euganea* Squinabol (1904).

Localities. CAS loc. 1144, Marca shale (MBP), Moreno Gulch, and Cima Hill hole III (at approximately 356·5 and 360 ft.).

Holotype. USNM 157968, N37/0 from upper Maestrichtian Moreno formation, Moreno Gulch (3517), Fresno County, California.

? *Sciadiocapsa causia* sp. nov.

Plate 7, figs. 2*a*, *b*

Description. Shell of three segments and a poorly developed velum; small hemispherical cephalis, hemispherical thorax, and flared discoidal abdomen. Cephalis with small, short, apical horn and small, short, upward-directed tube; surface generally covered with depressions and sometimes a few pores. Internally, weak paired ridges extending from the apical spine to the sides of the vertical spine are variably developed. There are six collar pores and, in some specimens, a tiny axial spine. Externally collar stricture well-defined. Thorax circular to slightly elliptical in transverse section, with largest dimension generally neither in the sagittal nor lateral plane. Circular to subcircular pores or depressions over all; small, closely spaced, and irregularly arranged proximally; larger and in transverse rows distally; frequently almost entirely filled by delicate meshwork. Tiny dorsal and lateral wings sometimes present. Lumbar stricture marked by almost 90° change in contour from thoracic wall; on some specimens, a few protrusions are present here on the inner wall between the pores, and a few specimens have a fragmentary velum. Circular to elliptical abdominal pores or depressions, varying greatly in size, are arranged in one to eight circular (transverse) rows. In some specimens the abdomen is almost completely perforated by pores and in others only one row at the lumbar stricture is present. Outer edge upturned, may be smooth or scalloped as from an incomplete row of pores.

Length of cephalis 25 μ, of thorax 35–50 μ; diameter of thorax 55–85 μ, of abdomen 125–230 μ. Dimensions based on twenty-seven specimens.

Discussion. This species differs from ? *S. petasus* in having pores not arranged in longitudinal rows, and only weak apical-vertical spine ridges.

Localities. CAS loc. 1144, Marca shale (MBP), Moreno Gulch, CAS loc. 39545, and UCMP loc. A2615.

Type specimens. Holotype USNM 157969, N37/0; paratype USNM 157970, N37/0, both from upper Maestrichtian Moreno formation, Moreno Gulch (3518), Fresno County, California.

? *Sciadiocapsa petasus* sp. nov.

Plate 7, figs. 1*a*, *b*

Description. Shell of three segments and poorly developed velum: small, lobed, hemispherical cephalis; hemispherical thorax; and flared discoidal abdomen. Cephalis with small, short, apical horn and upward-directed tube of moderate size; surface with minute papillae. Internally, strong paired ridges extend from the apical spine to the sides of the vertical spine. Externally, furrows along these ridges constrict the cephalis forming three lobes. Six collar pores are present, and externally the collar stricture is well defined. Thorax circular to slightly elliptical in transverse section, with largest dimension not uniformly in sagittal nor lateral plane. Proximal pores few, circular, and scattered; remainder circular to subangular, sometimes subdivided by meshwork, arranged in transverse and vertical rows. Tiny dorsal and lateral wings rarely present. Lumbar stricture marked by almost 90° change in contour from thoracic wall. Internally, some specimens have fragments of a velum with small circular pores here. Abdominal pores, circular to quadrangular, arranged in transverse and longitudinal rows, become uniformly larger distally or are very variable in size. The longitudinal rows, eleven to twelve per half a circumference at the lumbar stricture, are collinear with rows on the thorax. Distal edge slightly upturned and always ragged.

Length of cephalis 25–30 μ, of thorax 35–50 μ; diameter of thorax 70–85 μ, of abdomen 175–265 μ. Dimensions based on ten specimens.

Discussion. This species is distinguished from ? *S. causia* as described under that species. Specimens from CAS loc. 39545 have tube and horn considerably larger than upper Maestrichtian specimens.

Localities. CAS loc. 1144, Moreno Gulch, and CAS loc. 39545.

Holotype. USNM 157971, M37/4 from upper Maestrichtian Moreno formation, Moreno Gulch (3518), Fresno County, California.

Genus *Clathrocyclas* Haeckel 1881

Remarks. The following species are assigned only doubtfully to this genus because of their possessing a cephalic tube and a ragged distal margin instead of numerous radial apophyses.

? *Clathrocyclas diceros* sp. nov.

Plate 5, fig. 4

Description. Shell of three segments: cephalis hemispherical; thorax long, campanulate; abdomen flared, fragmentary. Cephalis with sparse, small, circular pores; slender upward-directed tube; and two basally connected horns. Apical horn slender, smooth except for slight roughness at tip and curved line at base; and dorsal horn lamellar, smooth, pointed, and short. Internally, weak paired ridges extend along cephalic wall from apical spine to sides of vertical spine. There are six collar pores with jugular pores extremely small, and a slender short axial spine. Thorax, with tiny primary lateral and dorsal wings, is elliptical to circular in transverse section; when elliptical there is no constant relation between the greatest width and the sagittal plane of the cephalis. Pores subcircular to subangular, small, irregularly arranged immediately below collar stricture, becoming larger distally, and arranged quincuncially in six to eight transverse rows. There are generally two weak internal ridges above and below the last row of pores; the first cannot always be distinguished and it is the second, situated where the wall flares, which is considered as the lumbar stricture. This stricture slightly raised dorsally. Abdomen shorter dorsally, more flared, sometimes almost horizontal, with smaller and fewer pores than ventrally; few specimens with almost no flare. Pores vary greatly in size and arrangement; some specimens with proximal pores similar to those of thorax becoming smaller distally; others with smaller, irregularly arranged pores over all. Distal margin ragged.

Length of cephalis 25 μ, of thorax ventrally 90–125 μ, dorsally 85–120 μ, of apical horn 30–45 μ, of dorsal horn 15–25 μ; diameter of thorax in widest plane 90–105 μ, in narrowest plane 65–90 μ. Dimensions based on sixteen specimens.

Discussion. This species differs from ? *Clathrocyclas lepta* as described under that species, from *Clathrocyclas* (*Clathrocyclia*) *tintinnaeformis* Campbell and Clark (1944b) in having a less campanulate thorax and a dorsal horn, and from *Clathrocyclas irrasa* Eicher (1960) in having a shorter thorax without a distinct basal ridge or septum.

Localities. CAS loc. 1144, Marca shale (MBP), Moreno Gulch, Cima Hill hole III (at approximately 306·5, 310, and 320 ft.), CAS loc. 39545, and Stanford loc. PL2875.

Holotype. USNM 157972, M38/2 from upper Maestrichtian Moreno formation, Marca shale (MBP), Fresno County, California.

? *Clathrocyclas lepta* sp. nov.

Plate 5, figs. 5*a, b*

Description. Shell of two segments; cephalis hemispherical, and thorax long, conical proximally, flared distally. Cephalis poreless with a slender, horizontal or slightly downward-directed tube, and one sturdy, three-bladed apical horn. One specimen of the ten examined had a second, small, rough horn lateral to the apical horn. Internally, some specimens have weak paired ridges extending from the apical to the secondary lateral spines. There are six collar pores, the jugular pores extremely small, and a tiny axial spine. Two sturdy, long, three-bladed spines (wings) extend outward from the secondary lateral spines at the collar stricture, and tiny dorsal and primary lateral wings extend from the thorax proximally. Proximal thoracic pores variable in size, small, subcircular to elliptical, and irregularly arranged, some with tiny prongs on their margin which occasionally join to form a fine meshwork. Distally they become rapidly larger, subcircular to subangular, arranged quincuncially in seven to eight transverse rows. Thoracic wall constricts slightly, approximately where the character of the pores changes, and flares at about the fifth or sixth row beyond; no constricting internal shelf at either of these levels. Distal edge ragged.

Length of cephalis 30–5 μ, of longest thorax 150 μ, horn 45–60 μ, of longest secondary lateral spine wing (all broken) 70 μ; diameter of thorax at first change of contour 50–65 μ. Dimensions based on ten specimens.

Discussion. This species differs from ? *C. diceros* in having sturdy secondary lateral spine wings, and in lacking a dorsal horn and internal ridges to define a lumbar stricture.

Locality. Moreno Gulch.

Holotype. USNM 157973, U26/4 from upper Maestrichtian Moreno formation, Moreno Gulch (3518), Fresno County, California.

? *Clathrocyclas hyronia* sp. nov.

Plate 5, figs. 1*a, b*

Description. Shell sturdy, of at least four segments; cephalis hemispherical, thorax campanulate, first abdominal segment annular, and fourth segment fragmentary, constricting. Cephalis with large downward-directed tube, sturdy, basally-ridged, apical horn with a rough blunt tip, and small, subcircular, scattered pores. Six collar pores with jugular pores extremely small. Two short, sturdy, ridged secondary lateral spines extend at the collar stricture, and the primary lateral and dorsal spines extend as small wings on the proximal thorax. Proximally, thoracic pores subcircular, small, of variable size and irregular arrangement; distally, larger, more uniform in size, arranged quincuncially and tending to form transverse rows. Prongs on their margin join to form a delicate meshwork more prominent proximally. Intervening bars have large blunt nodes around each pore. Lumbar stricture marked by an indentation externally and narrow sturdy ridge internally. First abdominal segment short with one to three rows of pores, similar in size and shape to those of the distal thorax, but without prongs and meshwork, and only rarely with surrounding nodes; separated from the next segment, which has only been observed as a ragged fragment with similar pores, by a stricture similar to, but not quite as heavy as and not always as uniformly developed as, the lumbar stricture.

Length of cephalis 30–5 μ, of thorax 95–140 μ, of abdomen 25–50 μ; diameter of thorax proximally 75–85 μ, distally 120–40 μ. Dimensions based on eleven specimens.

Discussion. This species differs from the other two species of ? *Clathrocyclas* described herein in the character of the thorax which is much heavier, its lower margin defined by a well-marked ridge; and the large downward-directed tube.

Localities. Marca shale (MBP) and Moreno Gulch.

Holotype. USNM 157974, N37/0 from upper Maestrichtian Moreno formation, Moreno Gulch (3518), Fresno County, California.

Genus *Theocampe* Haeckel 1887 emend. Burma 1959

Remarks. Of the genera and subgenera which Burma (1959) synonymized under this name, only *Theocamptra* has a type species in which a cephalic tube is indicated in the species description or illustration. However, it is suspected that the presence of the cephalic tube has been overlooked in a large number of described species.

It is difficult to compare many of the earlier described species of artostrobiids with these Cretaceous forms because of this circumstance, and also because the cephalis and thorax were frequently considered as one segment and the abdomen as divided into numerous segments by apparent strictures between the transverse rows of pores, so that it is not always certain how many segments are present. When topotypic material could not be examined, the relationships between species described here and those in previous literature were evaluated on the basis of surface features and size.

Perhaps after further study this genus will be subdivided on the basis of the character of the internal lumbar stricture which in the first three species described (*T. daseia, T. lispa,* and *T. bassilis*) is a well-developed shelf similar to that found in many cyrtoids. The remaining four species (*T. vanderhoofi, T. argyris, T. dactylica,* and *T. altamontensis*) differ in having the internal lumbar stricture defined only by a slight thickening of the shell wall.

Theocampe daseia sp. nov.

Plate 6, figs. 9a, b

Description. Shell of three segments; cephalis and thorax together conical with thick wall, and abdomen subcylindrical, thinner-walled, elliptical in transverse section. Cephalis bears a horizontal or slightly upward-directed tube and a sturdy, short, apical horn, ridged at the base. Cephalic pores sparse, small, circular. Internally, weak paired ridges extend from apical spine to sides of vertical spine. The thick cephalic wall obscures the internal dorsal features, but there is some evidence for the presence of a dorsal branch and paired ridges from the apical spine to the secondary lateral spines. There are six collar pores and an additional pair of pores ventrally to them in the lower wall of the tube. Axial spines, up to five in number (longest 68 μ) are present. Thoracic pores are circular, directed outward and downward, set in irregular circular to angular frames proximally and in angular to rounded arches distally; the most distal tend to form a transverse row above lumbar stricture, which is defined by a change in contour externally and a sturdy shelf internally. Surface of thorax, and sometimes the cephalis, with numerous minute nodes. The smooth abdomen viewed laterally is of constant width proximally and narrows distally to a thin-walled poreless peristome with smooth margin; viewed ventrally or dorsally it generally narrows almost immediately below the lumbar stricture. Abdominal pores, subcircular to elliptical, in from three to perhaps seven approximately evenly spaced, transverse rows; first row immediately below lumbar stricture.

Length of cephalis 20–5 μ, of thorax 35–45 μ (most 40 μ), of abdomen 70–95 μ (six complete abdomens measured), of apical spine 10–20 μ; width in sagittal plane of thorax 65–70 μ, of abdomen 63–75 μ; lateral width of thorax 58–65 μ, of abdomen 52–60 μ. Dimensions based on fifteen specimens.

Discussion. This species differs from *Theocampe bassilis*, in having a larger thorax and a distinct, sharp apical horn.

Localities. CAS loc. 1144, Marca shale (MBP), Moreno Gulch, Cima Hill hole III (at approximately 310 and 327·5 ft.), and CAS loc. 39545.

Type specimens. Holotype USNM 157980, M39/0 Moreno Gulch (3517); paratype USNM 157981, M39/4 CAS loc. 1144, both from upper Maestrichtian Moreno formation, Fresno County, California.

Theocampe lispa sp. nov.

Plate 6, fig. 11

Description. Shell of three segments, smooth, thin, hyaline, spindle shaped; cephalis and thorax together conical to truncate globose; abdomen medianly expanded. Thorax and abdomen both elliptical in transverse section with largest dimension in sagittal plane. Cephalis bears a horizontal or slightly downward-directed tube, and rarely a tiny apical horn; sparse pores small, circular. Internally, weak paired ridges extend from apical spine to sides of vertical spine. There are six collar pores and an additional pair of pores ventrally to them in the lower wall of the tube. Long (to 50 μ) curved axial spines, up to five in number, extend into the abdomen. Thoracic pores circular, widely spaced, tending to form a transverse row immediately above the lumbar stricture; this stricture defined by a change in contour externally and a sturdy shelf internally. Abdominal pores, circular to elliptical, in from four to seven, perhaps more, transverse rows. Proximally and distally these rows sometimes tend to merge irregularly. Abdominal wall thins distally to a smooth edge.

Length of two complete specimens 110 and 115 μ, of cephalis 20–2 μ, of thorax 23–8 μ, of abdomen 65 and 80 μ; width in sagittal plane of thorax 50–5 μ, of abdomen

65–75 μ, of aperture 35 and 40 μ; lateral width of thorax 40–5 μ, of abdomen 45–50 μ. Dimensions based on ten specimens.

Discussion. This species differs from *T. bassilis* as described under that species and from *Dictyocephalis* (*Dictyoprora*) *miralestensis* Campbell and Clark (1944a) in its smoother surface and fewer rows of pores on the abdomen.

Localities. CAS loc. 1144, Marca shale (MBP), and Moreno Gulch.

Holotype. USNM 157982, N41/1 from upper Maestrichtian Moreno formation, CAS loc. 1144, Fresno County, California.

Theocampe bassilis sp. nov.

Plate 6, fig. 10

Description. Shell of three segments, thin, hyaline; cephalis and thorax together conical; abdomen long, subcylindrical. Both thorax and abdomen elliptical in transverse section with largest dimension in sagittal plane. Cephalis bears a horizontal tube and in most specimens a tiny apical horn; pores sparse, small, circular. Internally, weak paired ridges extend from apical spine to sides of vertical spine. A dorsal branch from the apical spine, just below the apex, extends to the cephalic wall. There are six collar pores and an additional pair of pores ventrally to them on the lower wall of the tube. Axial spines, possibly five in number, are very variable in size, some extend into the abdomen (longest 70 μ). Thorax with minute nodes and rarely ridges over all; circular pores widely spaced, those above lumbar stricture forming a transverse row. Lumbar stricture well-defined, externally by a change in contour and internally by a distinct shelf. Abdomen with subcircular to subquadrangular pores, generally with prongs on their margin, in up to ten, perhaps more, transverse rows; the first immediately below the

---

EXPLANATION OF PLATE 6

Fig. 1. ?*Bisphaerocephalina apimelos* sp. nov., holotype USNM 157993, right lateral view.
Fig. 2. *Eribotrys litos* sp. nov., holotype USNM 157992, left lateral view.
Fig. 3. *Eribotrys despoena* sp. nov., holotype USNM 157990, right lateral view.
Fig. 4. *Eribotrys anax* sp. nov., holotype USNM 157991, right lateral view.
Fig. 5. *Rhopalosyringium sparnon* sp. nov., holotype USNM 157978, left lateral view.
Fig. 6. *Rhopalosyringium colpodes* sp. nov., holotype USNM 157979, ventral-left lateral view.
Figs. 7a, b. *Rhopalosyringium* ?*magnificum* Campbell and Clark. *a*, USNM 157975, right lateral view. *b*, USNM 157976, right lateral view.
Fig. 8. *Rhopalosyringium elasson* sp. nov., holotype USNM 157977, left lateral view.
Figs. 9a, b. *Theocampe daseia* sp. nov. *a*, holotype USNM 157980, left lateral view. *b*, paratype USNM 157981, right lateral view.
Fig. 10. *Theocampe bassilis* sp. nov., holotype USNM 157983, right lateral view.
Fig. 11. *Theocampe lispa* sp. nov., holotype USNM 157982, left lateral view.
Fig. 12. *Theocampe vanderhoofi* Campbell and Clark, USNM 157984, right lateral view.
Fig. 13. *Theocampe argyris* sp. nov., holotype USNM 157985, left lateral view.
Figs. 14a, b. *Theocampe altamontensis* (Campbell and Clark). *a*, USNM 157988, left lateral view. *b* USNM 157987, left lateral view.
Fig. 15. *Theocampe dactylica* sp. nov., holotype USNM 157986, right lateral view.
Fig. 16. ?*Theocampe stathmepora* sp. nov., holotype USNM 157989, right lateral view.
All figures ×251.

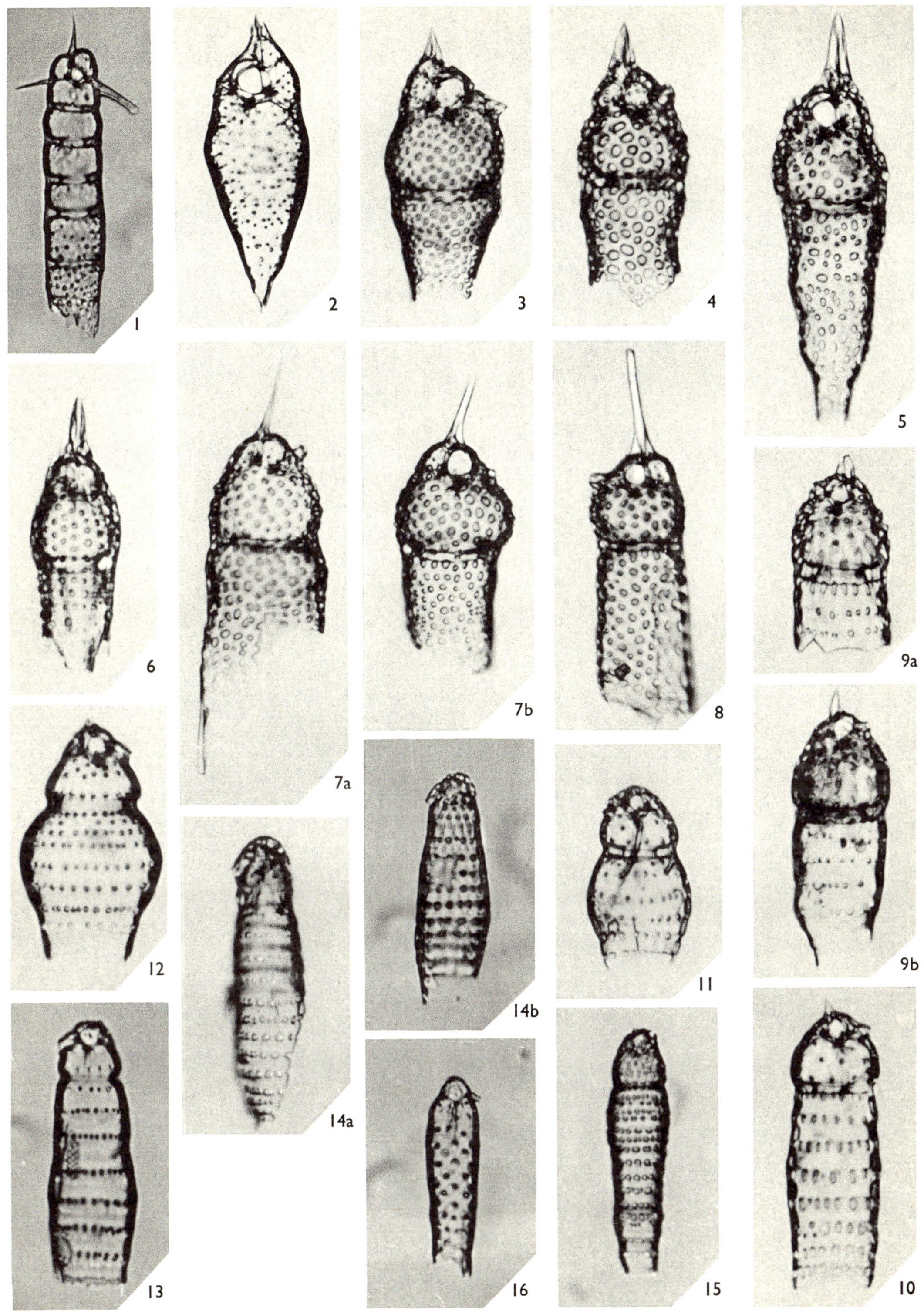

FOREMAN, Upper Maestrichtian Radiolaria of California

lumbar stricture. Rows approximately evenly spaced, except distally where the shell thins, constricts slightly, and rows are more closely spaced. Distal margin smooth.

Length of cephalis 20–3 μ, of thorax 25–30 μ, of longest broken abdomen 135 μ: of one complete abdomen 125 μ; width in sagittal plane of thorax 52–60 μ, of abdomen 58–70 μ; lateral width of thorax 45–58 μ, of abdomen 45–60 μ. Dimensions based on ten specimens.

Discussion. This species differs from *T. lispa* by the shape of the abdomen and the prongs on the margin of the abdominal pores, and from *T. daseia* as described under that species.

Localities. CAS loc. 1144, Marca shale (MBP), and Moreno Gulch.

Holotype. USNM 157983, M38/4 from upper Maestrichtian Moreno formation, Marca shale (MBP), Fresno County, California.

Theocampe vanderhoofi Campbell and Clark emend.

Plate 6, fig. 12

1944b *Theocampe (Theocampana) vanderhoofi* Campbell and Clark p. 34, pl. 7, fig. 19.

Description. Shell of three segments, spindle-shaped: cephalis and thorax together short, conical; abdomen long, medianly inflated. Both thorax and abdomen elliptical in transverse section with largest dimension generally in sagittal plane. Hyaline shell varies considerably in thickness among specimens. Cephalis bears a large downward-directed tube, and a small apical horn which varies from a short slender spine to a blunt thickening of an already thick cephalic wall; circular pores sparse. Internally, weak paired ridges extend from apical spine to sides of vertical spine. A dorsal branch from the apical spine near the apex extends to the cephalic wall, and weak paired ridges extend from the apical spine to the secondary lateral spines originating approximately at the point where this dorsal branch is attached. There are six collar pores and an additional pair of pores ventrally to them on the lower wall of the tube. Axial spines, when they can be distinguished, are short, up to perhaps five in number. The thorax proximally has irregular, small, circular pores, and distally a single row of circular, downward-directed pores immediately above the lumbar stricture. This stricture marked externally by a distinct indentation, but internally only by a thickening of the shell wall. Abdomen with six to twelve transverse rows, of elongated, subcircular to subquadrangular pores. Pores of first row, immediately below lumbar stricture, tend to be more circular, and proximal and distal rows are more closely spaced than median ones. Widest part of the abdomen generally at the fourth or fifth row rarely at the third or sixth row. Distally, abdominal wall thins and transverse rows become more irregular. Most distal portion poreless, narrowed, cylindrical to subcylindrical; margin smooth.

Length of cephalis 20–3 μ, of thorax 20–30 μ (most 25 μ), of complete abdomen 90–155 μ; width in sagittal plane of thorax 50–65 μ (most 55 μ), of abdomen 60–100 μ; lateral width of thorax 40–55 μ (most 45–50 μ), of abdomen 65–85 μ; distance between third and fourth row of abdominal pores 8–22 μ (most 10–15 μ), between fourth and fifth row 10–20 μ (most 15–20 μ). Dimensions based on forty specimens.

Discussion. This species differs from *T. argyris* as described under that species. It is the most common of the many artostrobiid species seen in the Maestrichtian of California.

Localities. CAS loc. 1144, Marca shale (MBP), Moreno Gulch, Cima Hill hole III (at approximately 306·5, 310, 313·5, 320, 327·5, 330, 340, 356·5, 360 ft.), CAS loc. 39545, UCMP loc. A2615, Stanford loc. PL2875, Evitt slide AE28, and Lamont V–18–129.

Illustrated specimen. USNM 157984, L38/1 from upper Maestrichtian Moreno formation, Moreno Gulch (3517), Fresno County, California.

Theocampe argyris sp. nov.

Plate 6, fig. 13; text-fig. 1, figs. 1*a–c*

Description. Smooth, thin, hyaline shell of three segments: cephalis and thorax together short, conical; abdomen long, subcylindrical, rarely slightly expanded medianly. Both thorax and abdomen elliptical in transverse section with largest dimension in sagittal plane. Cephalis bears a large downward-directed tube and rarely a tiny apical horn; pores sparse, small, circular. Internally, weak paired ridges extend from apical spine to sides of vertical spine and from apical spine to secondary lateral spines. A dorsal branch from the apical spine near the apex extends to the cephalic wall. There are six collar pores and an additional pair of pores ventrally to them in the lower wall of the tube. Up to five axial spines are present; these generally short, not extending beyond thorax. Thorax has proximally a few scattered circular pores and distally a single row of circular, downward-directed pores immediately above the lumbar stricture. Lumbar stricture marked externally by a distinct indentation and internally by a slight thickening of the shell wall. On the abdomen, immediately below the lumbar stricture, a row of fairly uniform, small, circular to elliptical, evenly spaced pores; there are from three to seven, perhaps more, additional rows, widely and approximately evenly spaced with pores subcircular to subquadrangular, more elongate than in the first row. Lower margin of most specimens ragged. Some specimens, however, have fragments of a thin-walled, slightly constricted, smooth peristome.

Length of cephalis 15–20 μ, of thorax 20–5 μ, of three abdomens 80 μ, 113 μ, and 120 μ, longest broken abdomen 145 μ; width in sagittal plane of thorax 40–7 μ, of abdomen 55–70 μ; lateral width of thorax 35–45 μ, of abdomen 40–55 μ; distance between third and fourth row of abdominal pores 15–30 μ (most 18–25 μ), between fourth and fifth row 13–25 μ (most 18–23 μ). Dimensions based on twenty specimens.

Discussion. This species is distinguished from *T. vanderhoofi* by its slender abdomen and fewer more widely spaced rows of abdominal pores, from *Tricolocampe* aff. *cylindrica* Nakaseko (1963) by its shorter more numerous axial spines and few scattered pores on the thorax proximally, and from *Lithocampe* (*Lithocampium*) *modeloensis* var. *longu* Campbell and Clark (1944*a*), by its lack of 'sinuous, simple or bifurcated, subvertical ribs'.

Localities. CAS loc. 1144, Marca shale (MBP), Moreno Gulch, and Cima Hill hole III (at approximately 360 ft.).

Holotype. USNM 157985, N37/0 from upper Maestrichtian Moreno formation, Moreno Gulch (3517), Fresno County, California.

Theocampe dactylica sp. nov.

Plate 6, fig. 15

Description. Shell very slender, spindle-shaped, of three segments. Cephalis with upward, almost vertical tube and small, circular, irregularly spaced pores in weak irregular frames. Internally, weak paired ridges extend from apical spine to sides of vertical spine. Six collar pores present, and axial spines, up to five on one specimen (longest 75 μ). Thoracic pores small, circular, set in downward-directed frames, arranged quincuncially. Lumbar stricture marked by slight indentation and thickening of shell wall. Sub-cylindrical abdomen very variable; generally slightly expanded proximally, tapering gently and sometimes unevenly to a smooth, poreless peristome of very variable width. Abdominal pores subcircular to subquadrangular, arranged in irregularly spaced, transverse rows, seven and fifteen on two complete specimens.

Length of cephalis 15–17 μ, of thorax 20–2 μ, of two complete abdomens 85 and 160 μ; diameter of thorax 30–5 μ, of abdomen 38–52 μ (most 40–2 μ); distance between third and fourth row of pores 5–8 μ, between fourth and fifth row of pores 5–10 μ. Dimensions based on fourteen specimens.

Discussion. This species differs from *T. altamontensis* (Campbell and Clark) (1944*b*) in having an upward-directed tube and pores generally smaller, in more closely spaced rows.

Localities. CAS loc. 1144, Marca shale (MBP), Moreno Gulch, CAS loc. 39545, Stanford loc. PL2875, and Evitt slide AE28.

Holotype. USNM 157986, N37/4 from upper Maestrichtian Moreno formation, Moreno Gulch (3517), Fresno County, California.

Theocampe altamontensis (Campbell and Clark)

Plate 6, figs. 14*a, b*

1944*b* *Tricolocampe* (*Tricolocamptra*) *altamontensis* Campbell and Clark p. 33, pl. 7, figs. 24, 26.

Description. Shell very slender, spindle-shaped, of three segments. Cephalis with a large downward-directed tube. Deep, circular to angular, closely spaced frames enclose most of the small circular pores, forming a honeycomb effect over most of the cephalis. Internal structures cannot be seen clearly, but there are indications that six collar pores are present; as many as eight long, curved, axial spines have been counted in one specimen (longest 105 μ). Externally, no or very little change in contour at either collar or lumbar stricture. Both thorax and abdomen defined by the distinct change in character of their pores. Thoracic pores small, circular, set in downward-directed frames, arranged quincuncially in two or three transverse rows. A broad poreless band marks the lumbar stricture; internally there is only a slight thickening of the wall. Subcylindrical abdomen slightly expanded medianly. Abdominal pores subcircular to subquadrangular, generally with prongs on their margin medianly, arranged in up to seventeen, perhaps more, transverse rows. These rows approximately equidistant except distally where they are more closely spaced. All specimens observed had a ragged lower margin.

Length of cephalis 20–3 μ, of thorax 20–2 μ, of longest incomplete abdomen 155 μ; diameter of thorax 30–5 μ; greatest width of abdomen 50–62 μ; distance between third and fourth row of abdominal pores 10–18 μ, between fourth and fifth row 10–18 μ. Dimensions based on seven specimens.

Discussion. This species differs from *T. dactylica* as described under that species.

Localities. CAS loc. 1144, Marca shale (MBP), Moreno Gulch, UCMP loc. A2615, CAS loc. 39545, and Stanford loc. PL2875.

Illustrated specimens. USNM 157987, M38/0 Moreno Gulch (3517), and USNM 157988, K38/0 CAS loc. 1144, both from upper Maestrichtian Moreno formation, Fresno County, California.

? Theocampe stathmepora sp. nov.

Plate 6, fig. 16

Description. Shell of two segments, very slender, subcylindrical, with minute papillae over all. Cephalis small, dome-like, poreless, a tiny apical horn and a small, slender, downward-directed tube. Internally, weak paired ridges extend from apical spine to sides of vertical spine, a tiny dorsal branch extends from apical spine to dorsal cephalic wall immediately below apex, and six collar pores are present. There are as many as seven, or perhaps more, long (up to 60 μ), sometimes curved, axial spines. The slender thorax is very slightly expanded medianly and narrows slightly distally, where the wall thins and its lower margin is ragged. Circular to subcircular thoracic pores set in weak angular frames are quincuncially arranged in diagonal rows. Pores are small proximally, gradually increasing in size distally, or increasing in size at a point about one-fifth to one-quarter down the length of the thorax with a slight change in contour at this point (rarely sufficiently distinct to give the appearance of a segmental division).

Length of cephalis 18 μ, of longest thorax 130 μ; diameter of thorax 35–45 μ. Dimensions based on eleven specimens.

Discussion. This species is probably more common than indicated in the chart but because of its small size is frequently lost in preparation. It differs from *Theocampe altamontensis* in having only two segments and pores arranged in diagonal rows.

Although this species is very closely related to others of the genus *Theocampe* it is only doubtfully assigned because of the lack of a clearly defined lumbar stricture and the pores in diagonal rows.

Localities. CAS loc. 1144, Marca shale (MBP), and Moreno Gulch.

Holotype. USNM 157989, N37/0 from upper Maestrichtian Moreno formation, Moreno Gulch (3517), Fresno County, California.

Genus Rhopalosyringium Campbell and Clark 1944b emend.

Diagnosis. Tricyrtid forms, with cephalis having a tube, apical horn, and internally a dorsal branch from the apical spine immediately below the top of the cephalic wall not forming a dorsal lobe. Pores of thorax and abdomen not regularly transversely aligned.

Remarks. This genus is distinguished from *Eribotrys* by the absence of an apical-dorsal cephalic lobe. The presence or absence of a dorsal foot is here regarded as a distinguishing feature at the specific rather than at the generic level.

Rhopalosyringium ? *magnificum* Campbell and Clark

Plate 6, figs. *7a, b*

? 1944*b* *Rhopalosyringium magnificum* Campbell and Clark, p. 30, pl. 7, figs. 16, 17.

Description. Shell of three segments: cephalis hemispherical, thorax globose, and abdomen cylindrical proximally, narrowing irregularly distally. Cephalis with horizontal or slightly upward-directed tube and a long, slender apical horn, smooth except for its ridged base. Pores sparse, scattered, subcircular, most of them dorsal. Internally, paired ridges extend from apical spine to sides of vertical spine, and from apical spine to secondary lateral spines. A dorsal branch, very close to the apex, extends to the cephalic wall. Six collar pores present and two additional paired pores ventrally in the lower wall of the tube. A short axial spine can sometimes be seen. Thorax with small dorsal and lateral wings variably present. Thoracic pores vary from fairly uniform in size, subcircular shape, and distribution, to more irregular; either type with or without marginal prongs or subdividing meshwork. Intervening bars may be smooth, or with nodes or weak angular frames. Lumbar stricture well-defined by a change of contour, with smooth abdomen generally narrower than thorax, and elliptical in transverse section. Pores circular to elliptical in shape and irregularly distributed. A single, long, slender, dorsal foot (never seen unbroken) is incorporated as a rib in the abdominal wall for at least one-third of the abdominal length and extends downward from the distal, thin, ragged margin.

Length of cephalis 25–30 μ, of thorax 35–50 μ (most 40–5 μ), of abdomen 60–95 μ, of longest broken horn 125 μ, of longest broken foot 75 μ; diameter of thorax 65–92 μ, of abdomen 55–72 μ. Dimensions based on twenty-two specimens.

Discussion. This species differs from *R. elasson* as discussed under that species. It is only doubtfully considered conspecific with *R. magnificum* Campbell and Clark (1944*b*) because the latter differs slightly in having abdominal pores arranged in crudely transverse rows and apical horn ridged.

Localities. CAS loc. 1144, Marca shale (MBP), Moreno Gulch, Cima Hill hole III (at approximately 313·5 and 340 ft.), CAS loc. 39545, and Evitt slide AE26.

Illustrated specimens. USNM 157975, M38/0 and USNM 157976, N35/3, both from upper Maestrichtian Moreno formation, Moreno Gulch (3518), Fresno County, California.

Rhopalosyringium elasson sp. nov.

Plate 6, fig. 8

Description. Shell of three segments: cephalis hemispherical; thorax short, cylindrical to annular; abdomen long, cylindrical, sometimes expanded slightly proximally. Cephalis with horizontal tube and long, slender, smooth (except for ridged base) apical horn, with sparse, small, subcircular pores, especially dorsally. Internally, paired ridges extend from apical spine to sides of vertical spine, and from apical spine to secondary lateral spines, and a dorsal branch, very close to the apex, extends to the cephalic wall. Six collar pores, and two additional paired pores ventrally in the lower wall of the tube, and in at least one specimen a short axial spine. Thoracic pores irregular in size and distribution, circular to subcircular, occasionally with a few marginal prongs which sometimes join to subdivide them. Rarely a small dorsal wing is present. Lumbar stricture marked by very little or no change in contour. Abdominal pores similar to those of thorax. A long, slender, smooth, dorsal foot, incorporated in the abdominal wall as a rib for a considerable distance, and never observed unbroken, extends from the lower ragged margin.

Length of cephalis 25–8 μ, of thorax 25–30 μ, of abdomen 50–133 μ, of longest broken horn 85 μ, of longest broken foot 65 μ; diameter of thorax 45–65 μ, of abdomen 45–70 μ. Dimensions based on ten specimens.

Discussion. This species is distinguished from *R. ?magnificum* Campbell and Clark (1944*b*) by its smaller thorax with less regular pores.

Localities. CAS loc. 1144, Marca shale (MBP), Moreno Gulch, and CAS loc. 39545.

Holotype. USNM 157977, M38/1 from upper Maestrichtian Moreno formation, Marca shale (MBP), Fresno County, California.

Rhopalosyringium sparnon sp. nov.

Plate 6, fig. 5

Description. Shell of three segments; cephalis hemispherical, thorax globose, abdomen subcylindrical to inverted conical, narrowly cylindrical distally. Cephalis with sturdy, ridged apical horn; two ridges from this horn extend on to cephalis ventrally, and an upward-directed tube which does not protrude emerges between them. Sparse, small, circular pores, sometimes in weak angular frames, are mostly dorsal. Internally, paired ridges extend from apical spine to sides of vertical spine and from apical spine to secondary lateral spines. A short dorsal branch extends to the cephalic wall just below the apex. It is not always complete and is so close to the cephalic wall that it does not form a dorsal lobe. Six collar pores and a short axial spine are present. Pores of the thorax are very variable in size, subdivided by a fine meshwork, and set in weak angular frames. Lumbar stricture well-defined by a change in contour on those specimens with a subcylindrical abdomen, and to a lesser extent on those specimens with an inverted conical abdomen. Abdominal pores subcircular to elliptical, proximally with prongs on their margin or subdivided by meshwork. Pores of some specimens tend toward transverse alignment.

Length of cephalis 30–5 μ, of thorax 40–55 μ, of abdomen 125–85 μ, of horn 37–50 μ; diameter of thorax 70–95 μ, of widest part of abdomen 65–80 μ, of aperture 12–25 μ. Dimensions based on seven specimens.

Discussion. This species differs from *R. colpodes* in lacking longitudinal ridges.

Localities. CAS loc. 1144, Marca shale (MBP), and Moreno Gulch.

Holotype. USNM 157978, P39/0 from upper Maestrichtian Moreno formation, CAS loc. 1144, Fresno County, California.

Rhopalosyringium colpodes sp. nov.

Plate 6, fig. 6

Description. Shell of three to ? four segments, spindle-shaped. Cephalis with a sturdy, broad-based, ridged horn; ridges bifurcating and extending throughout length of shell. An upward-directed tube that does not protrude emerges between ridges which are more prominent at this point. Circular pores are few, scattered. Internally, paired ridges extend from apical spine to sides of vertical spine, and from apical spine to secondary lateral spines and a dorsal branch, very near the apex, extends to the cephalic wall. Six collar pores and a short axial spine are present. Thorax annular to subglobose, with approximately sixteen to twenty ridges on the circumference. Pores, tending to form longitudinal rows between these ridges, are variable in size, circular to elliptical, occasionally with prongs on their margin which sometimes join to subdivide them. Little or no change in contour at the lumbar stricture. Abdomen cylindrical proximally with pores similar to those of thorax, tending to form transverse as well as longitudinal rows; distally the shell wall constricts, thins and has a ragged margin, and pores become smaller, more widely spaced.

Length of cephalis 30–5 μ, of thorax 30–45 μ, of abdomen with ragged margin 50–125 μ, of horn 25–60 μ, of axial spine 15–25 μ; diameter of thorax 50–75 μ, of abdomen 50–70 μ. Dimensions based on nine specimens.

Discussion. This species differs from *R. sparnon* as described under that species. One specimen was observed with the third segment slightly constricted internally almost forming a fourth segment.

Localities. CAS loc. 1144, Moreno Gulch, and Cima Hill hole III (at approximately 356·5 ft.).

Holotype. USNM 157979, P37/3 from upper Maestrichtian Moreno formation, CAS loc. 1144, Fresno County, California.

Genus *Eribotrys* gen. nov.

Type species. *E. despoena* sp. nov.

Diagnosis. Shell of two or three segments, cephalis with the three usual cephalic lobes and a fourth apical-dorsal lobe separated from the dorsal lobe by a dorsal branch from the apical spine. Septum between cephalic and dorsal lobe with one set of paired pores.

This genus is distinguished from all other cyrtellarians in possessing an apical-dorsal cephalic lobe, its lower margin a dorsal branch from the apical spine.

Remarks. Members of this genus have some basic differences from Recent cannobotryids and some fundamental similarities with them. Details of the structure of the cephalic chambers of Recent cannobotryids have been discussed by several authors but here we use as a basis for comparison the results of Petrushevskaya (1964, 1965), which are the most comprehensive, and Bütschli (1892).

There has in the past been some confusion in terminology due to different authors considering opposite faces of the radiolarian cephalis as anterior and posterior. Petrushevskaya (1964, 1965), following Bütschli (1892), considered that portion of the cephalis associated with the dorsal spine as anterior ('antecephalic chamber') and that associated with the vertical spine as posterior ('postcephalic chamber'). Haeckel (1887), on the other hand, considered the face associated with the dorsal spine as dorsal and the opposite face associated with the vertical spine as frontal or ventral; Nigrini (1967) also used dorsal and ventral in the sense of Haeckel. Therefore in order to avoid the ambiguity of using the term 'antecephalic' for that part of the cephalis associated with the dorsal spine and 'postcephalic' for the opposite face the terms dorsal and ventral in the sense of Haeckel are used here.

In the Cretaceous forms at least, the cephalis is not so subdivided as to form completely separate chambers, and it seems best to regard all the area above the collar pores as a cephalis divided into: a cephalic lobe over the cardinal and cervical pores, a ventral lobe ventral to the cephalic lobe (including at least the proximal part of the tube and with an additional set of sometimes poorly defined pores basally), a dorsal lobe over the jugular pores, and an apical-dorsal lobe above the dorsal lobe. These lobes are homologous with Petrushevskaya's cephalic chamber, postcephalic chamber, lower part of antecephalic chamber, and upper part of antecephalic chamber. In the Cretaceous forms the two paired ridges which extend along the inner cephalic wall from the apical spine to the sides of the vertical spine make the wall of the cephalic lobe when viewed laterally appear heavier than it actually is. *Eribotrys* differs from Tertiary and Recent forms in that the apical portion of the dorsal lobe is separated off by a dorsal branch from the apical spine to form the apical-dorsal lobe. It differs also in that the paired median branches of Bütschli and Petrushevskaya, which divide the septum separating the cephalic and dorsal lobe to form two sets of paired pores, are lacking and therefore the dividing septum is broken by only one set of paired pores. This septum in the Cretaceous specimens is formed by the longitudinal apical spine (A), basally the secondary lateral spines (l), and apically the lamellar ridge (LR) at the apical-dorsal edge of the cephalic lobe (text-fig. 1, fig. 11*e*).

Bütschli's drawings for *Lithobotrys geminata* (1882, pl. 8, figs. 27*a–c*), reproduced herein (text-fig. 1, figs. 11*a–c*), are somewhat ambiguous as in his key and fig. 27*a* (11*a*) (basal view of collar pores) he indicates that *e* represents the primary lateral spines, while in his fig. 27*c* (11*c*) (dorsal view) he shows his secondary lateral spines labelled *e*. If one accepts the designation in fig. 27*c* (11*c*) of *e* as primary lateral spines, one could interpret the basal set of paired pores as oblique cardinal pores and therefore the dividing septum with only one set of paired pores would be homologous with the dividing septum of *Eribotrys* (compare figs. 11*c* and 11*e*, text-fig. 1). However, examination of specimens of *Lithobotrys geminata* from Barbados shows that this interpretation would be incorrect and that *L. geminata* does indeed have a dividing septum with two sets of paired pores. Petrushevskaya (1965, fig. 1, III) apparently recognized this ambiguity in Bütschli's drawings, but in so doing has relabelled Bütschli's correctly labelled fig. 27*a*, (11*a*) to conform to his incorrectly labelled fig. 27*c* (11*c*), so that in her drawings *e* represents the secondary lateral spines.

It is not certain whether the lamellar ridge (LR) of the Cretaceous forms is homologous with the lamellar ridge which is the upper margin of the dividing septum between the cephalic and dorsal lobes of the Eocene and Recent forms of Bütschli and Petrushevskaya (l of Bütschli) or with their paired median branches which divide the septum (h of Bütschli). It appears that the latter alternative is much more likely and that, at least in the Eocene forms, the vague ridges which extend from the median branches along the cephalic wall towards or to the vertical spine may be homologous with the more distinct apical-vertical ridges of the Cretaceous forms.

? *Lithomelissa heros* and *Acidnomelos apteron* exhibit some fundamental similarities to these Cretaceous cannobotryids in that they have a large cephalis with six collar pores, and paired ridges extending on the inner cephalic wall from the apical spine to the sides of the vertical spine and from the apical spine to the secondary lateral spines. These ridges, however, are not constantly present, and the apical-lateral spine ridges are much less developed than in the cannobotryids.

The artostrobiids and the members of the genus *Rhopalosyringium* described herein show even more similarities. In addition to the large cephalis, and the two pairs of ridges from the apical spine to the vertical spine and sometimes to the secondary lateral spines, there is a dorsal branch from the apical spine to the cephalic wall. It is, however, so close to the apex that it does not form a dorsal-cephalic lobe. Also there are two paired pores ventral to the collar pores forming the proximal part of the lower wall of the tube when it is outward- or downward-directed.

Eribotrys despoena

Plate 6, fig. 3; text-fig. 1, figs. 9*a*, *b*

Description. Shell of three segments, spindle-shaped. Cephalis lobed as typical of this genus, thorax annular, inflated, and abdomen cylindrical proximally, constricted distally to a narrowly circular aperture. Cephalis with a large downward-directed tube, and a small, sharp, ridged horn. Small scattered pores are more prominent dorsally and on the proximal part of the tube. Cephalic lobe of cephalis with few pores and frequently tiny nodes. Internally, besides the dorsal branch and the apical-secondary lateral septum, paired ridges extend from the apical spine to the sides of the vertical spine. An axial spine and six collar pores are present with an additional pair of pores ventrally to them on the lower wall of the tube. Thoracic pores, circular to subcircular, sometimes with prongs or subdividing meshwork, rarely in angular frames, are fairly uniformly distributed. Lumbar stricture well-defined internally, but with only a slight change or no change in contour externally. Proximally, abdominal pores similar to those of thorax, with a few marginal prongs, distally becoming smaller and tending to form transverse rows.

Length of cephalis (measured from top of apical-dorsal lobe) 30–40 μ, of thorax 35–55 μ, of one complete abdomen 95 μ; width in sagittal plane of thorax 65–85 μ, of abdomen 62–85 μ. Dimensions based on eight specimens.

Discussion. This species differs from *Eribotrys anax* as described under that species.

Localities. CAS loc. 1144, Marca shale (MBP), Moreno Gulch, CAS loc. 39545, and UCMP loc. A2615.

Holotype. USNM 157990, M38/2 from upper Maestrichtian Moreno formation, CAS loc. 1144, Fresno County, California.

Eribotrys anax sp. nov.

Plate 6, fig. 4

Description. Shell of three segments, spindle-shaped. Cephalis divided into cephalic, ventral, dorsal, and small apical-dorsal lobes, with a short, sharp, broad-based, ridged apical horn and a horizontal to slightly upward-directed tube. Internally, besides the dorsal branch and the apical-secondary lateral septum there are strong paired ridges from the apical spine to the sides of the vertical spine, and a single branch from the vertical spine downward to the ventral wall separating the proximal tube from the thorax. Six collar pores and a long axial spine are present. Cephalic pores, subcircular and irregularly spaced are more numerous dorsally. Two specimens from Moreno Gulch had short secondary lateral spines protruding at the collar stricture. Thorax annular to subglobose with pores fairly uniformly distributed, variable in size and shape, generally subdivided by a delicate meshwork and surrounded by weak angular frames. Dorsal and primary lateral wings small, poorly developed, and variably present. Lumbar stricture well-defined internally but with little change in contour externally. Subcylindrical abdomen narrows irregularly distally to a ragged margin. Pores subcircular to elliptical, irregularly arranged, and frequently subdivided by a delicate meshwork, especially proximally.

Length (exclusive of horn) of longest specimen 210 μ, of horn 10–30 μ, of cephalis (measured from top of apical-dorsal lobe) 32–40 μ, of thorax 35–50 μ, of longest abdomen 125 μ, of axial spine, up to 80 μ; width of thorax 60–85 μ, of abdomen 62–90 μ. Dimensions based on twelve specimens.

Discussion. This species differs from *E. despoena* in having a horizontal to upward-directed tube.

Localities. CAS loc. 1144, Marca shale (MBP), Moreno Gulch, Cima Hill hole III (at approximately 310 and 320 ft.), and CAS loc. 39545.

Holotype. USNM 157991, M38/4 from upper Maestrichtian Moreno formation, Moreno Gulch (3517), Fresno County, California.

Eribotrys litos sp. nov.

Plate 6, fig. 2; text-fig. 1, figs. 10*a, b*

Description. Shell of two segments, spindle-shaped, elliptical in transverse section. Cephalis divided into cephalic, ventral, dorsal, and apical-dorsal lobes with a short, sharp, ridged apical horn and a horizontal tube which does not protrude. Internally, besides the dorsal branch defining the lower margin of the apical-dorsal lobe and the apical-secondary lateral septum separating the cephalic and dorsal lobes, there are strong paired ridges from the apical spine to the sides of the vertical spine. These constrict the cephalic lobe so that viewed ventrally it appears to be divided into three parts. A downward-directed branch from the vertical spine just below the tube extends to the ventral wall (sometimes protruding slightly) and defines the lower margin of the ventral lobe. Six collar pores and a short (longest 20 μ) axial spine are present. Cephalic pores circular to subcircular, widely scattered, except that there are none or very few on the cephalic lobe where there are minute papillae on a few specimens. Subcylindrical thorax constricts to a narrow aperture distally. Pores circular to elliptical, occasionally with marginal prongs, widely and irregularly spaced, distally tending to transverse alignment.

Length (including horn and tube) 155–68 μ, of cephalis (measured from top of apical-dorsal lobe) 35–45 μ, of thorax 100–5 μ, of apical horn 10–20 μ; width of thorax in sagittal plane 52–70 μ (most 58–65 μ). Dimensions based on twelve specimens.

Discussion. This species differs from *E. anax* and *E. despoena* in possessing only two segments.

Localities. CAS loc. 1144, Marca shale (MBP), and Moreno Gulch.

Holotype. USNM 157992, N39/1 from upper Maestrichtian Moreno formation, CAS loc. 1144, Fresno County, California.

Genus Bisphaerocephalina Petrushevskaya 1965

? Bisphaerocephalina apimelos sp. nov.

Plate 6, fig. 1; text-fig. 1, figs. 8a, b

Description. Shell of seven to perhaps nine segments, long, slender, elliptical in transverse section. Cephalis poreless, covered with minute nodes, with a smooth, slender, short, apical horn and at the poorly defined collar stricture a long, slender, slightly downward-directed tube and a long, slender, dorsal spine. Internally, apical-secondary lateral septum with one set of paired pores, and strong paired ridges from approximately the median point on the apical spine to the sides of the vertical spine. Six collar pores. Thorax short, poreless; occasionally with two tiny primary lateral spine wings. Post-thoracic segments, all of approximately equal width, vary considerably in length with no regular increase in size except that generally the two distal segments are the longest. Strictures dividing the segments irregular in that they are sometimes neither complete nor horizontal. Proximal segments with very tiny papillae are generally poreless except for a few irregularly spaced inward- and downward-directed pores at some of the strictures. Distal segments, from approximately the fifth or seventh, commonly with small, circular to elliptical pores, irregularly arranged. Wall of last segment thins and constricts slightly, with ragged distal margin.

Length of five segments 90–130 μ, of eight segments on two complete specimens 205 and 225 μ, of cephalis 20–7 μ, of thorax 15–25 μ, of tube 25–30 μ, of horn 20–5 μ, of shortest abdominal segment 15–28 μ, of longest abdominal segment 25–40 μ; width in sagittal plane of cephalis 30–5 μ, of thorax 30–6 μ, of abdomen 35–45 μ. Dimensions based on twelve specimens.

Discussion. The cephalic structure in this species is fundamentally similar to that of *Eribotrys* except that there is no dorsal branch on the apical spine separating off an apical-dorsal lobe. Another difference is that the paired ridges between the apical and the vertical spines originate lower on the apical spine, and appear in lateral view to separate off the lower part of the cephalic lobe of the cephalis.

There seems little doubt that the similarities in general appearance, of the cephalis of this Cretaceous form and of the described Recent species of *Bisphaerocephalina*, reflect true homologies. It particularly resembles *B. armata* Petrushevskaya (1965, p. 93, fig. 8, III–V). It is assigned only doubtfully to *Bisphaerocephalina* because ? *B. apimelos* has only one set of paired pores in the dividing septum between cephalic and dorsal lobes, and it is at present not certain whether this, together with the numerous post-thoracic segments constitute an important enough difference to erect a separate genus.

? *B. apimelos* also differs from *B. armata* in the shape and length of tube, and absence of sturdy primary lateral spine wings.

Localities. CAS loc. 1144 and Moreno Gulch.

Holotype. USNM 157993, N37/4 from upper Maestrichtian Moreno formation, Moreno Gulch (3517), Fresno County, California.

Genus *Lithocampe* Ehrenberg 1838

Remarks. Although the species described below has a small apical horn and shows no evidence of being constricted distally, and in these respects does not conform to the definition of *Lithocampe*, it nevertheless appears more similar to the type species of that genus than to the type species of *Lithomitra* Bütschli (1882) or that of *Eucyrtidium* Ehrenberg (1847a). The evidence for relationship to *Lithocampe* is however not sufficient to justify emendation of the generic definition.

? *Lithocampe eureia* sp. nov.

Plate 8, fig. 6

Description. Shell of five to six segments; cephalis and thorax together conical, abdominal segments cylindrical, frequently elliptical in transverse section. Cephalis small, with short, slender, rough horn and occasionally a tiny spine extending horizontally or downward from the vertical spine. Pores sparse, small, circular; surface roughened by tiny thorns. Internally, paired ridges extend from apical spine to sides of vertical spine and from apical spine to secondary lateral spines. Six collar pores present and collar stricture well-defined. Thoracic pores subcircular, generally closely spaced, arranged quincuncially when uniform in size, irregularly when less uniform. Tiny thorns or weak angular frames roughen the surface. Abdominal segments all with pores variable in size arranged in rather uneven transverse rows, four to six rows on first abdominal segment, three to six on second and third. Proximal segments have tiny thorns or weak angular frames and pores subcircular to elliptical, while distal segments have a smoother surface with pores tending to be quadrangular. Strictures between fourth, fifth, and sixth segments frequently interrupted, irregular or rarely lacking. When this is the case the fourth or fifth segment is longer than indicated in the dimensions. Sixth segment observed only as a fragment with up to three rows of pores.

Length of cephalis 18–20 μ, of apical horn 7–15 μ, of thorax 22–36 μ (most 30–4 μ) of first abdominal segment 20–35 μ (most 28–32 μ) of second abdominal segment 18–35 μ (most 25–30 μ) of third abdominal segment 12–32 μ; over-all (exclusive of horn) of four segments 85–113 μ, of five segments 125–140 μ; greatest width of thorax 45–58 μ, of first abdominal segment 52–65 μ (most 60–5 μ), of second abdominal segment 57–80 μ (most 60–70 μ). Dimensions based on twenty specimens.

Discussion. This species is distinguished by its small size and abdominal segments with pores very variable in size arranged in closely spaced transverse rows. Also, it differs from *Stichopilium annulatum* Popofsky (1913) in having more cylindrical segments and more pronounced abdominal strictures, and from *Lithocampe* sp. Nigrini (1967) in the latter character.

Localities. CAS loc. 1144, Marca shale (MBP), Moreno Gulch, Cima Hill hole III (at approximately 363·5 ft.), and CAS loc. 39545.

Holotype. USNM 157973, U26/0 from upper Maestrichtian Moreno formation, Moreno Gulch (3518), Fresno County, California.

Genus *Dictyomitra* Zittel 1876

?1903 *Diplostrobus* Squinabol 1903, p. 140.

Remarks. Although no material from the type locality for *Dictyomitra multicostata* Zittel (the type species) was available, a small sample and a picked slide from A. Elger, from Oberg bei Peine, near Zittel's type locality have provided some specimens (Pl. 7, figs. 4*a*, *b*) which correspond very well with Zittel's original description. Oberg bei Peine (Quadratenkreide) is a Campanian locality and Zittel's locality Vordorf (Mukronatenkreide) a Maestrichtian one (Gignoux 1955, p. 422). Among the specimens from Oberg bei Peine are a number with a definitely constricted last segment. Also one of Zittel's illustrations (1876, pl. 2, fig. 4) shows a fragment of the last preserved segment which begins to constrict. It therefore seems that Zittel's original description of *Dictyomitra* was based on incomplete material and this together with other evidence indicates that the generic definition should be revised. This step is, however, deferred until topotypic material is available for study.

The species below are assigned to *Dictyomitra* because of their apparent similarity to the illustrations of the type specimens in that they all have a small simple cephalis, relatively small thorax, longitudinal ribs, and distalmost segments which are cylindrical or show some degree of constriction.

Dictyomitra cf. *multicostata* Zittel

Plate 7, figs. 9*a*, *b*

1876 *Dictyomitra multicostata* Zittel, p. 81, pl. 2, figs. 2, 3, 4.
1944b *Dictyomitra multicostata* Campbell and Clark, p. 39, pl. 8, fig. 42 (non figs. 22, 23, 24, 29, 35).
1944b *Lithocampe andersoni* var. *paucisepta* Campbell and Clark, p. 43, pl. 8, fig. 13.

Remarks. Striated forms belonging to the genus *Dictyomitra* are very variably developed in the material studied. Six species are described. Because topotypic material of *D. multicostata* is not available and because there is very real doubt that the species described here is conspecific with *D. multicostata s. s.* the name is here not certainly assigned to the species described below.

Description. Shell of seven to fifteen segments, conical except for last one or two segments which narrow; sometimes one or two segments cylindrical before narrowing begins. Cephalis without horn but thickened apically, small, poreless, with surface smooth except for costae. Internally, four collar pores. Thorax equal to, or shorter than, cephalis. Generally, postcephalic segments increase gradually in width and length; in some specimens, however, the distal segments are markedly longer than the proximal ones. Segments generally inflated, their greatest dimension medianly. Costae which begin on the cephalis extend longitudinally throughout (twelve to seventeen at the widest segment). Commonly they follow the contours of the inflated segments, but may also be straight-edged forming a smooth silhouette. A single transverse row of circular to elliptical pores (one pore, or rarely two, between adjacent costae) is present at or immediately above each stricture. Distally an additional row may be present, generally above, rarely below the stricture, and the distalmost segments frequently have pores in longitudinal rows, irregularly spaced. Proximal segments may be smooth except for the costae, or have single longitudinal rows of transversely elliptical depressions between costae. The last constricting segment is frequently lacking; when present it appears as if rather irregularly compressed with margin ragged.

Length of complete specimens of ten and twelve segments 250 and 310 μ, of eight segments 150–90 μ, of cephalis 12–15 μ, of thorax 10–12 μ, of widest segment 30–58 μ; width of widest segment 65–115 μ; distance between costae at widest segment 8–14 μ. Dimensions based on twenty specimens.

Discussion. D. striata Lipman (1952) may be conspecific with the species described above. Lipman distinguishes it from *D. multicostata* Zittel by size and lack of pores; however Zittel's illustrated specimen (pl. 2, fig. 2) 245 μ long is well within Lipman's size range and both Koslova and Gorbovetz (1966) and Aliev (1965) indicate that this species does have pores. Type material was not available for study.

D. multicostata has been reported from the following localities ranging in age from Senonian to Maestrichtian. In most cases the material, to judge from the lack of detail in drawings and photographs, was poorly preserved, and although there is a superficial resemblance to *D. multicostata*, the presence of a great variety of striated forms in the Cretaceous (Holmes 1900, Tan Sin Hok 1927, Aliev 1965) prevents reliable identifications on the basis of superficial similarity. At present, therefore, this species does not appear to be useful for precise age determinations.

1891 *Dictyomitra multicostata* Perner, J., p. 259, 265, pl. 10, fig. 1. From Priesen, layer 2 d (Senonian), also in Vunic, Střem, and Postelberg, Bohemia.

1892 *Dictyomitra multicostata* Rüst, D., p. 109, pl. 16, fig. 3. From Porcupine Mountains, Northwest Manitoba, Canada. Base of Pierre shale (Campanian).

1900 *Dictyomitra multicostata* Holmes, W. M., p. 710, pl. 38, fig. 3. From Upper Chalk at Coulsdon, Surrey, England (Senonian).

1903 *Dictyomitra multicostata* Squinabol, S., p. 139. Mentions common occurrence in siliceous nodules of Teolo, C. Brustolo and Monte Sereo, Italy. No description or illustration (Upper Cretaceous).

1925 *Dictyomitra multicostata* Richter, M., p. 557, pl. 9, fig. 17. From Cape Conway, Staten Island off Cape Horn (Upper Cretaceous).

1944b *Dictyomitra multicostata* Campbell and Clark, p. 39, pl. 8, fig. 42 (non figs. 22, 23, 24, 29, 35). From UCMP, loc. A2615, Mitchell Creek, Alameda County, California, U.S.A. 'Moreno Grande' formation (lower Maestrichtian–? upper Campanian, Goudkoff D–2).

1947 *Dictyomitra multicostata* Nauss, A. W., p. 341, pl. 48, figs. 3, ?8. From Vermilion area, east-central Alberta, Canada, upper 50 ft. of Lloydminster shale and lower half of Lea Park shale lower Campanian–upper Santonian. 'The species is widespread in occurrence throughout Alberta, Saskatchewan and Manitoba.'

1956 *Dictyomitra* (*D.*) cf. *D.* (*D.*) *multicostata* Zittel Bolin, E. J., p. 295, pl. 39, fig. 19. From Lyons County, Minnesota, U.S.A., Upper Niobrara formation or possible lower Pierre shale, (Coniacian–Santonian or possibly lower Campanian).

1963 *Dictyomitra* (*Dictyomitra*) *multicostata* Pessagno, p. 206, pl. 1, figs. 9, 10; pl. 4, fig. 3; pl. 5, figs. 7, 13. From Puerto Rico, Cariblanco formation and Parguera limestone (early Campanian).

Localities. CAS loc. 1144, Marca shale (MBP), Moreno Gulch, Cima Hill hole III (at approximately 275·5, 306·5, 310, 313·5, 320, 327·5, 330, 340, 356·5, 360, and 383 ft.), CAS loc. 39545, UCMP loc. A2615, Stanford loc. PL2875, Evitt slides, and Lamont V–18–129.

Illustrated specimens. USNM 157994, Q36/4, Q35/3 from upper Maestrichtian Moreno formation, CAS loc. 1144, Fresno County, California.

Dictyomitra lamellicostata sp. nov.

Plate 7, figs. 8*a*, *b*

Description. Shell of six to twelve segments, conical except for distalmost segments which are generally cylindrical. Cephalis without a horn but thickened apically, small, poreless, smooth, except for costae. Internally four collar pores. Thorax equal to or shorter than cephalis. Segments increase gradually in length throughout. Proximal segments inflated, distal segments more annular. Individual segments tend to be widest distally. Costae, which begin on the cephalis and extend longitudinally throughout, are straight edged to form a smooth silhouette, generally developed as wide lamellae proximally, and widely spaced distally (ten to fourteen at widest segment). Rare specimens have costae of which the outer edges follow the contour of the segments. A single transverse row of large circular to triangular pores is present at each stricture, one pore between adjacent costae. Distally an additional row may be present. Single longitudinal rows of transversely elliptical depressions between adjacent costae begin at about the third segment. At the aperture the costae extend as slightly inward-directed teeth with some lamellar shell material at either side. Generally the teeth are separate immediately below the last stricture, rarely lamellar shell material joins them proximally.

Length of complete specimens including teeth, of six segments 178 μ, of ten segments 317 μ, of longest specimen of twelve segments with broken teeth 361 μ, of eight segments 173–230 μ, of cephalis 10–15 μ, of thorax 10–12 μ, of last segment 30–58 μ; width of last segment 70–140 μ, distance between costae on widest segment 12–20 μ. Dimensions based on twenty specimens.

Discussion. The distinctive apertural teeth and the widely spaced and proximally broad lamellar costae distinguish this species from all other costate stichocyrtids. However, when the teeth are broken off and the proximal broad lamellar costae not well-developed, it cannot always be distinguished confidently from other forms.

Localities. CAS loc. 1144, Marca shale (MBP), Moreno Gulch, Cima Hill hole III (at approximately 323·5, 340, 360, and 383 ft.), and UCMP loc. A2615.

Type specimens. Holotype USNM 157994, R39/0 CAS loc. 1144; paratype USNM 157994, H35/0 Moreno Gulch (3518), both from upper Maestrichtian Moreno formation, Fresno County, California.

Dictyomitra rhadina sp. nov.

Plate 7, figs. 5*a*, *b*

Description. Shell of nine to nineteen segments, possibly more, long, slender, sub-cylindrical. Cephalis small, poreless, with few scattered depressions and thickened pointed apex; internally, four collar pores. Longitudinal costae begin on the short thorax, and extend throughout length of shell. Generally, postcephalic segments increase very gradually in width to a postmedian segment, after which shell may be cylindrical for a segment or two and then narrow slightly. Last segment irregularly compressed with distal margin ragged. Segments usually increase somewhat irregularly in length, the longest segment generally median. Proximal segments have longitudinal rows of shallow depressions between costae. A single transverse row of circular pores, one pore or rarely two between adjacent costae, immediately below each stricture. The distalmost segments may also have other pores, irregularly distributed in single longitudinal rows between the costae.

Length of longest almost complete specimen of eighteen segments plus a fragment 356 μ, of shortest almost complete specimen of eight segments plus a fragment 163 μ, of eight segments 150–80 μ, of cephalis 15 μ, of thorax 10–12 μ, of fourth segment 20–7 μ, of eighth segment 20–30 μ, of longest segment 26–35 μ; width of thorax 25–30 μ, of fourth segment 35–40 μ, of eighth segment 46–58 μ; number of costae on half of eighth segment 11–14; distance between costae on eighth segment 5–8 μ. Dimensions based on fifteen specimens.

Discussion. This species differs from *D.* cf. *multicostata* by its small size and slender, cylindrical shape. It is probably more common than appears from this study as the small diameter may cause its loss during preparation.

EXPLANATION OF PLATE 7

Figs. 1*a*, *b*. ? *Sciadiocapsa petasus* sp. nov., holotype USNM 157971. *a*, apical view of cephalis. *b*, apical view of thorax and abdomen.

Figs. 2*a*, *b*. ? *Sciadiocapsa causia* sp. nov. *a*, holotype USNM 157969, apical view of abdomen. *b*, paratype USNM 157970, right lateral view.

Fig. 3. ? *Sciadiocapsa ptesimolecis* sp. nov., holotype USNM 157968, apical view.

Figs. 4*a*, *b*. *Dictyomitra multicostata* Zittel, USNM 158006, from Oberg bei Peine, Niedersachsen, Germany.

Figs. 5*a*, *b*. *Dictyomitra rhadina* sp. nov. *a*, holotype USNM 157995. *b*, paratype USNM 157996.

Figs. 6*a*–*d*. *Dictyomitra andersoni* (Campbell and Clark). *a*, *b*, USNM 157999; *a*, M38/0; *b*, M38/2, *c*, USNM 158001. *d*, USNM 158000.

Figs. 7*a*, *b*. *Dictyomitra* cf. *crassispina* (Squinabol). *a*, USNM 157997. *b*, USNM 157998.

Figs. 8*a*, *b*. *Dictyomitra lamellicostata* sp. nov. *a*, paratype USNM 157994, R39/0. *b*, holotype USNM 157994, H35/0.

Figs. 9*a*, *b*. *Dictyomitra* cf. *multicostata* Zittel. *a*, *b*, USNM 157994; *a*, Q35/3; *b*, Q36/4.

All figures ×166, except Fig. 3, ×251.

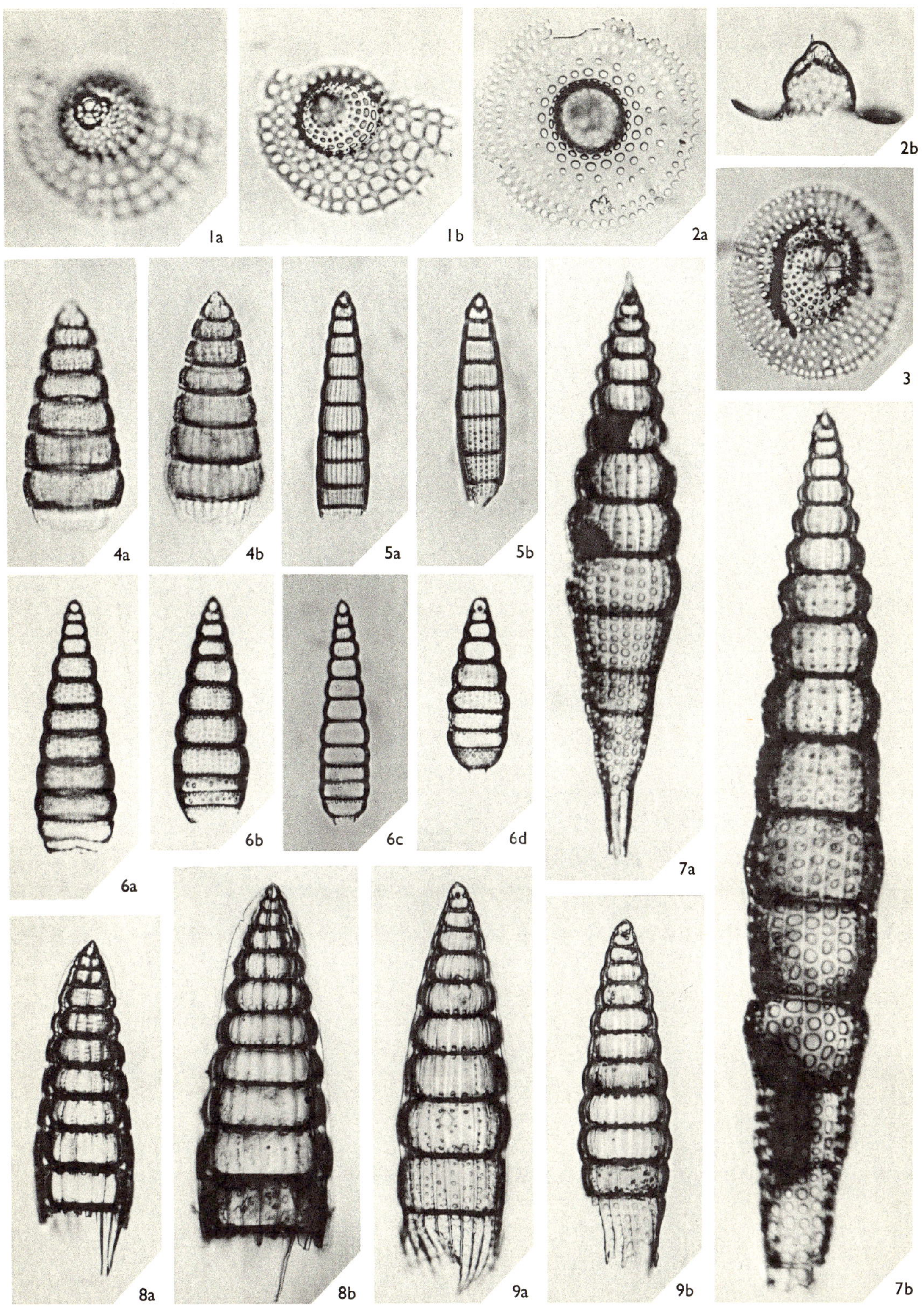

FOREMAN, Upper Maestrichtian Radiolaria of California

Localities. CAS loc. 1144, Marca shale (MBP), Moreno Gulch, CAS loc. 39545, and Stanford loc. PL2875.

Type specimens. Holotype USNM 157995, M38/0; paratype USNM 157996, L37/2, both from upper Maestrichtian Moreno formation, Moreno Gulch (3517), Fresno County, California.

Dictyomitra cf. *crassispina* (Squinabol)

Plate 7, figs. 7*a, b*

1903   *Diplostrobus crassispina* Squinabol, p. 140, pl. 8, fig. 37.

Description. Shell of nine to at least seventeen segments, spindle-shaped. Cephalis small, simple, poreless, smooth with short apical horn which varies considerably in shape, length, and width, rarely vestigial. Thorax equal to, or shorter than cephalis. Conical proximal segments, seven to twelve in number, increase gradually in length and width to approximately the median point, with abdominal segments (except sometimes the first) inflated and strictures deep. In profile, inflated segments tend to be widest on proximal half of each segment. Distal one to five segments inversely conical, not inflated with no or very little indentation at strictures, generally not continuing to increase regularly in length, the last constricting rapidly to form a long slender tube. Longitudinal costae extend from second or third segment throughout the length of the shell. The proximal conical part of the shell with a single transverse row of circular pores, one pore between adjacent costae, at each stricture, and a single longitudinal row of depressions between the costae. These occasionally broken by widely spaced, small circular pores medianly. The inversely conical distal part with circular pores; these becoming larger, more irregular in arrangement and shape, sometimes almost quadrangular on the distalmost segments.

Length (exclusive of apical horn) of eight complete specimens 270–545 μ, of longest broken specimen of sixteen segments plus a fragment 840 μ, of eight segments 155–232 μ, of horn 5–20 μ, of cephalis 12–15 μ, of thorax 10–15 μ; width of thorax 25–30 μ, of eighth segment 80–120 μ, of widest segment 75–125 μ, of tube 12–20 μ. Dimensions based on twenty-two specimens not only from CAS loc. 1144, Marca shale (MBP), and Moreno Gulch, but also from a fourth, Cima Hill hole III at approximately 360 ft.

Discussion. This species is distinguished from *D.* cf. *multicostata* by its apical horn, markedly inflated conical segments with greatest width proximally, long slender apertural tube, and lack of costae on the cephalis. Because of the small illustration and brief description it is impossible confidently to synonymise this species with *D. crassispina* Squinabol until topotypic material is available.

Localities. CAS loc. 1144, Marca shale (MBP), Moreno Gulch and Cima Hill hole III (at approximately 310, 320, 327·5, 340, and 360 ft.).

Illustrated specimens. USNM 157997, M38/3 and USNM 157998, M40/0, both from upper Maestrichtian Moreno formation, Marca shale (MBP), Fresno County, California.

Dictyomitra andersoni (Campbell and Clark) emend.

Plate 7, figs. *6a–d*

1944*b* Lithocampe (*Lithocampanula*) *andersoni* Campbell and Clark, p. 42, pl. 8, fig. 25.
1944*b* Dictyomitra (*Dictyomitroma*) *multicostata* Zittel; Campbell and Clark, p. 39, pl. 8, figs. 22–4, 29, 35, (non 42).
1944*b* Lithomitra (*Lithomitrissa*) *regina* var. *subconica* Campbell and Clark, p. 41, pl. 8, fig. 28.
?1944*b* Dictyomitra (*Dictyomitroma*) *tiara* Campbell and Clark, p. 40, pl. 8, figs. 1, 2, 3, 4, 12; homonym, non *D. tiara* Holmes 1900, p. 701, pl. 38, fig. 4.

Description. Shell of seven to fifteen segments, possibly more; spindle-shaped to conical with constricted aperture. Cephalis small, simple, frequently with a short, slender, apical horn and occasionally a tiny vertical horn; surface smooth, poreless; four collar pores. Thorax small, equal to or shorter than cephalis. Postcephalic segments annular, slightly inflated, either conical proximally and subcylindrical distally or conical throughout; except for last one or two segments which always narrow. Segments do not increase regularly in length; longest segments frequently in median section and distalmost with shorter segments between. Except for the single row of small circular pores at the strictures, surface ornamentation and presence or absence of other pores and depressions very variable. Generally, small circular depressions in longitudinal rows begin on the thorax or first abdominal segment, and extend for approximately two-thirds the shell length, the raised area (costae) between the rows wavy. Depressions occasionally pierce shell wall as pores; sometimes they are few or absent and then the ridges are straighter or lacking on some or all segments; when absent, segments smooth, hyaline. The distalmost segments smooth, poreless or with pores irregularly spaced, circular, elliptical, or triangular. Aperture with a short, smooth, poreless collar on well-preserved specimens.

Length (exclusive of horn and collar) of complete specimen of eight segments 160 μ, of twelve segments 280 μ, of longest broken specimen of fourteen segments plus a fragment 305 μ, of ten segments 160–257 μ, of cephalis 12–15 μ, of thorax 8–12 μ, of longest segment 25–40 μ; greatest width 58–95 μ. Dimensions based on twenty-three specimens from CAS loc. 1144, Marca shale (MBP), Moreno Gulch, UCMP loc. A2615 (the type locality), CAS loc. 39545, and Lamont V–18–129.

Discussion. This species differs from *Dictyomitra* cf. *multicostata* in the length of its segments which do not increase in size regularly, costae which do not extend on to the cephalis or the distalmost segments, and aperture with collar. It is the most common stichocyrtid in the upper Maestrichtian samples studied.

Localities. CAS loc. 1144, Marca shale (MBP), Moreno Gulch, Cima Hill hole III (at approximately 306·5, 310, 320, 327·5, 330, 356·5, and 360 ft.), UCMP loc. A2615, CAS loc. 39545, Stanford loc. PL2875, Evitt slides, Lamont V–18–129, and Trinidad well Marac 1.

Illustrated specimens. USNM 157999, M38/0, M38/2 and USNM 158000, M38/0, both from upper Maestrichtian Moreno formation, Moreno Gulch (3517), Fresno County, California; and USNM 158001, N38/2 from lower Maestrichtian– ? upper Campanian 'Moreno Grande' formation, CAS loc. 39545, Alameda County, California.

Dictyomitra regina (Campbell and Clark) emend.

Plate 8, figs. *5a–c*

1944*b* Lithomitra (*Lithomitrissa*) *regina* Campbell and Clark, p. 41, pl. 8, figs. 30, 38, 40.
?1904 *Dictyomitra crebrisulcata* Squinabol, p. 231, pl. 10, fig. 1.

Description. Shell of six to fourteen segments, possibly more, slender, conical to cylindrical with blunt apex. Cephalis small, cephalis and thorax together conical without distinct external collar stricture, partly depressed in thorax; four collar pores. Pores of cephalis small, circular, irregularly arranged. Postcephalic segments increase gradually in length and width except for the one or two distalmost segments which generally narrow slightly. Pores circular to elliptical, approximately equal in size throughout or increasing slightly in size distally; when elliptical their long axes transverse. Pores on proximal segments frequently in circular frames, those of distal segments occasionally almost quadrangular in shape. Costae extend from the cephalis throughout length of shell. They separate longitudinal rows of pores which may or may not be transversely aligned, with no break in the longitudinal spacing at the strictures. Distal margin generally ragged, or rarely with a ring of generally short constricting teeth formed by continuations of the costae.

Length of longest incomplete specimen of twelve segments plus a fragment 415 μ, of ten segments 250–325 μ, of cephalis 15–22 μ (most 15–18 μ), of cephalis and thorax together 27–43 μ (most 27–35 μ); width in sagittal plane of cephalis 15–25 μ (most 15–20 μ), of thorax 30–50 μ (most 40–5 μ), greatest width 65–120 μ. Dimensions based on thirty specimens measured not only from the Fresno County, California localities but also from CAS loc. 39545 and Lamont V–18–129.

Discussion. Some specimens (five of the thirty) measured have one or two of the median segments markedly elongated, and the segments immediately above and below correspondingly shortened. The elongated segments have less regular costae and pores, and are here considered aberrant. Specimens from CAS loc. 39545 tend to be shorter, entirely conical, with long sharp teeth distally.

This species may be conspecific with *Dictyomitra crebrisulcata* Squinabol (1904) but until topotypic material is available it cannot be confidently identified. It differs from *Stichophormis Montis Serei* Squinabol (1903) in lacking a horn and having a larger thorax and less distinct external strictures.

Localities. CAS loc. 1144, Marca shale (MBP), Moreno Gulch, CAS loc. 39545, UCMP loc. A2615, Stanford loc. PL2875, and Lamont V–18–129.

Illustrated specimens. USNM 158003, L38/3 CAS loc. 1144, USNM 158004, N40/3 Marca shale (MBP), and USNM 158005, O36/3 Marca shale (MBP), all from upper Maestrichtian Moreno formation, Fresno County, California.

? *Dictyomitra ascelis* sp. nov.

Plate 8, fig. 7

Description. Shell small, of six to seven or possibly more segments; cephalis and thorax together conical; abdominal segments subcylindrical, frequently elliptical in transverse section with greatest dimension in sagittal plane and little or no indentation at the strictures. Cephalis small, partly depressed in thorax, with no or very small blunt apical horn. Pores small, circular, irregularly arranged. Internally, four collar pores and paired ridges along cephalic wall from apical spine to sides of vertical spine. Postcephalic segments have small, circular, uniform pores increasing in size not at all, or only slightly distally. Thoracic pores irregular, or arranged in longitudinal rows; post-thoracic segments with pores in longitudinal rows, approximately thirteen to eighteen per half circumference on the fifth segment. Postcephalic segments increase gradually in size distally and the last observed segment constricts slightly to a ragged margin.

Length of longest specimen of seven segments 163 μ, of five segments 100–25 μ, of cephalis 15–17 μ, of cephalis and thorax together 27–32 μ, of fifth segment 25–40 μ; width in sagittal plane of thorax 38–50 μ, of fifth segment 55–65 μ. Dimensions based on fourteen specimens.

Discussion. This species is distinguished by its small size, pores in longitudinal rows, and lack of longitudinal costae. It differs from ? *L. regina* in its smaller size and lack of costae and from *Carpocanium conicum* Squinabol (1903) in being less conical with more numerous rows of pores. This species is only doubtfully assigned to *Dictyomitra* because of the presence of paired apical-vertical ridges in the cephalis.

Localities. CAS loc. 1144, Marca shale (MBP), Moreno Gulch, CAS loc. 39545, and UCMP loc. A2615.

Holotype. USNM 158002, M39/1 from upper Maestrichtian Moreno formation, Moreno Gulch (3517). Fresno County, California.

Genus *Stichopilidium* Haeckel 1887 emend.

Type species. Stichopilium macropterum Haeckel 1887, p. 1438 = ? *Rhopalocanium* sp. Bury 1862, pl. 17, fig. 7.

Remarks. The type species of *Stichopilium (Stichopilidium)*, *S. macropterum* Haeckel is similar to the species described here, in the shell being acutely conical with cephalis and thorax relatively small and with three flattened, broad-based, long wings.

As the shell of the type species of *Stichopilidium* seems to be of fundamentally different structure from that of the type species of *Stichopilium (Stichopilium)*, *Stichopilium bicorne* (Haeckel 1887, p. 1437, pl. 77, fig. 9) it is suggested that the former be used as a generic rather than a subgeneric name.

Stichopilidium teslaense (Campbell and Clark) emend.

Plate 8, fig. 13

1944b *Stichopilium (Stichopilidium) teslaense* Campbell and Clark, p. 36, pl. 7, fig. 49.

Description. Shell of five to eight, possibly more segments, conical except for last one or two cylindrical segments which generally narrow slightly. Cephalis poreless with a sturdy, rarely slender, smooth, long, conical horn; its base usually completely covering apex; and a vertical horn, generally small, but rarely almost as broad at its base as the apical horn. Surface roughened with tiny papillae. Internally, apical and vertical spines free, and weak paired ridges extend from apical spine along cephalic wall toward vertical spine. Two tiny jugular pores present in addition to the four larger collar pores, and there is a long, slender axial spine. Collar stricture marked externally by a change in contour. Thorax small with few scattered pores, surface roughened by nodes and irregular spines. Flattened, smooth, long, dorsal and primary lateral wings extend horizontally. Their bases generally spread over complete length of thorax and most of the first abdominal segment, rarely more slender with bases only on thorax. Post-thoracic segments increase regularly in size. Pores circular to subcircular, irregularly spaced, rather uniform, becoming more irregularly shaped and larger distally. Surface roughened by irregular spiny projections, except for last one or two cylindrical segments which are smoother with thinner walls. Distal margin never seen other than narrowed with ragged edge. Three to six, generally four, sturdy ribs originating, in most cases at about first or second abdominal segment, extend along length of shell and beyond to form free feet. They are incorporated in the shell wall proximally; free, attached by spiny structures distally; or incorporated in shell wall throughout; and are approximately evenly distributed around the shell with no particular relationship to the wings.

Length overall (exclusive of horn and feet) of longest specimen of eight segments 290 μ, of five segments 150–75 μ, of cephalis 15–22 μ, of thorax 20–5 μ, of longest broken wing 222 μ, of apical horn 45–80 μ; width of apical horn near base 5–20 μ, width of widest abdominal segment 90–130 μ. Dimensions based on fifteen specimens.

Discussion. The specimens studied agree well with Campbell and Clark's specimen except that the wings on their specimen are smaller. It differs from *Stichopilium bonum* Koslova (Koslova and Gorbovetz 1966) in having prominent longitudinal ribs and smaller more numerous abdominal pores irregularly arranged.

Localities. CAS loc. 1144, Marca shale (MBP), Moreno Gulch, Cima Hill hole III (at approximately 306·5 and 313·5 ft.), CAS loc. 39545, UCMP loc. A2615, Stanford loc. PL2875, and Lamont V–18-129.

Illustrated specimen. USNM 158007, M39/0 from upper Maestrichtian Moreno formation, CAS loc. 1144, Fresno County, California.

Genus *Stichomitra* Cayeux 1897

Remarks. The group of species here described under this generic name are very diverse, but further investigation is required to provide a practical basis for dividing them among several genera. There is an initial problem, in that no type species has yet been designated for *Stichomitra* but the species described below as belonging to this genus are characterized by numerous segments, small cephalis and thorax, and no pronounced longitudinal ribs or thoracic wings. The proximal part of the shell of these species is acutely conical and the distal part generally cylindrical or slightly constricted.

In addition there is in the upper Maestrichtian a large group of stichocyrtids with a cephalis having strong, paired apical-vertical ridges and a small apical horn; with median segments very variably developed and, in most cases, the distalmost segments broken off. This is an amorphous group and these incomplete specimens cannot confidently be identified. All complete specimens have been divided into three species on the basis of the character of their markedly different distalmost segments, and are here identified as ? *Stichomitra*. ? *S. alamedaensis* which has these ridges developed to a greater degree, so that they form a transverse ring, is also included.

Stichomitra compsa sp. nov.

Plate 8, figs. 8*a*, *b*

Description. Shell of six to eleven segments, conical except for last one or two segments which narrow. No complete specimen has been seen, but there is some indication that the last segment constricts to form a short tube. Cephalis with few scattered pores and tiny apical and vertical horns. On some specimens a dorsal branch from the apical spine forms a tiny dorsal horn. Internally, apical spine is free, and there are two tiny jugular pores in addition to the four larger collar pores. Weak paired ridges extend along the cephalic wall from apical spine to sides of vertical spine. Cephalis markedly elliptical in cross section, broad in lateral plane. Thoracic pores small, circular, irregularly spaced. Abdominal segments, except for the first, mostly uniform in length, or increase only very slightly distally. Pores vary considerably among specimens but are rather uniform on individual specimens. They vary from small, circular, widely spaced, very uniform, quincuncially arranged with little or no tendency towards transverse alignment, and surrounded by no or very low nodes or spines, and very rarely with some pores subdivided; to large, circular to subcircular, more closely spaced, less uniform, quincuncially arranged in transverse rows, and surrounded by blunt angular frames or nodes. Very rarely surface spines are longer, branch and join to form a partial rough lattice. Pores of adjoining segments generally apposed, sometimes coalesced. Median segments generally with three to five rows of pores.

Length of seven segments 150–75 μ, of longest broken specimen (ten segments plus fragment) 267 μ, of widest segment 25–40 μ; greatest width 70–110 μ. Dimensions based on twenty-five specimens.

Discussion. Variation in pore size and surface ornamentation is very great; as all degrees of intergradation are present it does not seem wise to subdivide this species on these

---

EXPLANATION OF PLATE 8

Fig. 1. ? *Stichomitra cechena* sp. nov., holotype USNM 158019, right lateral view.

Figs. 2*a*, *b*. ? *Stichomitra livermorensis* (Campbell and Clark). *a*, USNM 158018, right lateral view. *b*, USNM 158017, right lateral view.

Figs. 3*a–c*. ? *Stichomitra campi* (Campbell and Clark). *a*, USNM 158016, left lateral view. *b*, USNM 158014, right lateral view. *c*, USNM 158015, right lateral view.

Fig. 4. ? *Stichomitra alamedaensis* (Campbell and Clark), USNM 158020, left lateral view.

Figs. 5*a–c*. *Dictyomitra regina* (Campbell and Clark). *a*, USNM 158004, lateral view. *b*, USNM 158005, lateral view. *c*, USNM 158003, dorsal view.

Fig. 6. ? *Lithocampe eureia* sp. nov., holotype USNM 157973, U26/0, left lateral view.

Fig. 7. ? *Dictyomitra ascelis* sp. nov., holotype USNM 158002, left lateral view.

Figs. 8*a*, *b*. *Stichomitra compsa* sp. nov. *a*, holotype USNM 158008, lateral view. *b*, paratype USNM 158009, right lateral view.

Fig. 9. *Stichomitra cathara* sp. nov., holotype USNM 158010, left lateral view.

Figs. 10*a–c*. *Stichomitra asymbatos* sp. nov. *a*, paratype USNM 158013, left lateral view. *b*, holotype USNM 158011, dorsal view. *c*, paratype USNM 158012.

Fig. 11. *Amphipyndax plousios* sp. nov., holotype USNM 158024.

Figs. 12*a–c*. *Amphipyndax stocki* (Campbell and Clark). *a*, USNM 158021. *b*, USNM 158022. *c*, USNM 158023.

Fig. 13. *Stichopilidium teslaense* (Campbell and Clark), USNM 158007, right lateral view.

All figures × 166.

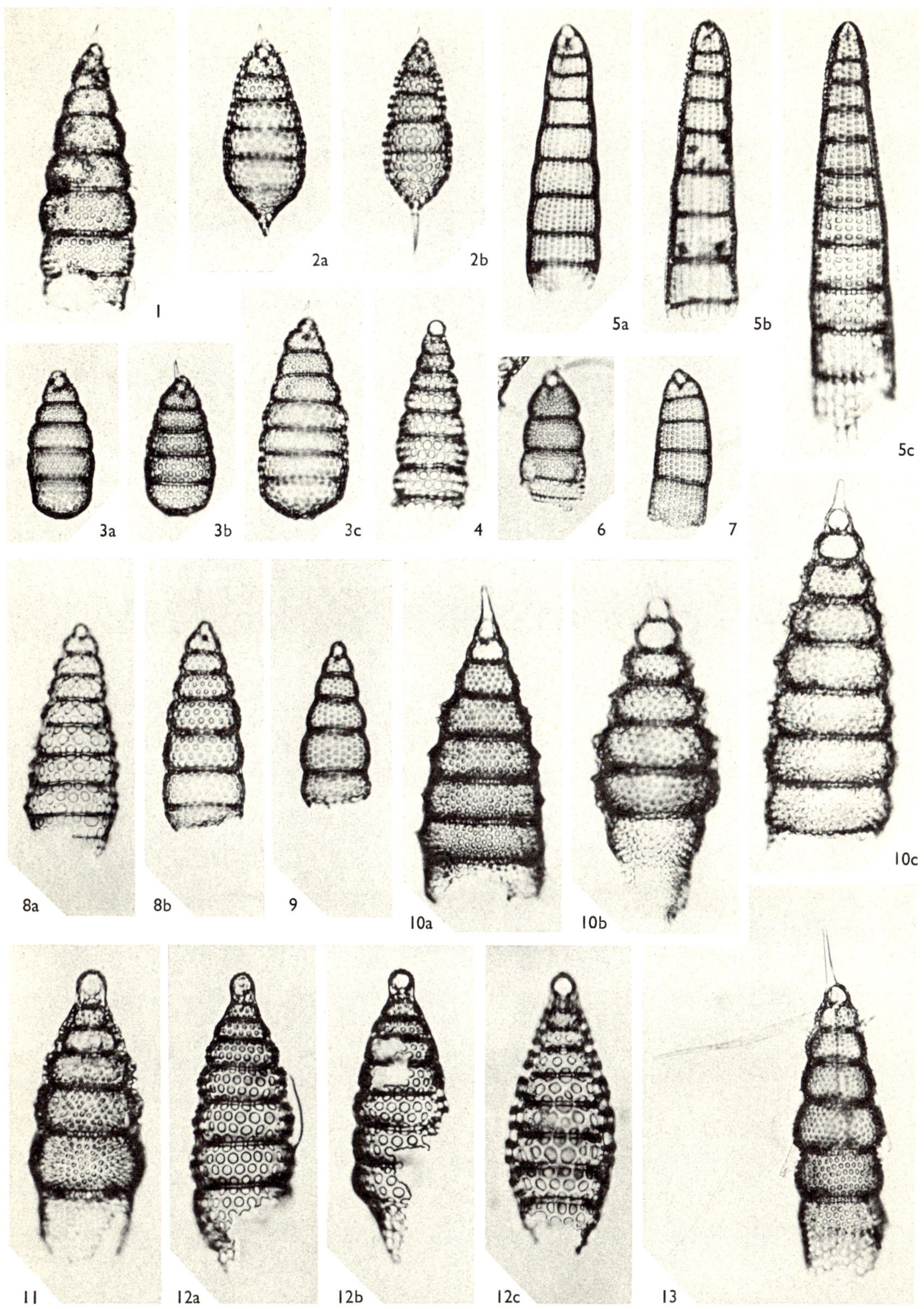

FOREMAN, Upper Maestrichtian Radiolaria of California

bases. This species differs from *Lithocampe obesa* Squinabol (1903) and from *Dictyomitra ferosia* Aliev var. *kelevudacika* Aliev (1965) in its smaller size.

Localities. CAS loc. 1144, Marca shale (MBP), Moreno Gulch, Cima Hill hole III (at approximately 310, 327·5, and 360 ft.), and CAS loc. 39545.

Type specimens. Holotype USNM 158008, O36/1 Marca shale (MBP); paratype USNM 158009, N38/3 Moreno Gulch (3517), both from upper Maestrichtian Moreno formation, Fresno County, California.

Stichomitra cathara sp. nov.

Plate 8, fig. 9

Description. Shell of six to seven segments, first five or six conical, last one or two cylindrical or narrowing slightly; distal margin broken, ragged, slightly turned in. Cephalis small with no or few scattered pores and tiny apical and vertical horns. Internally, weak paired ridges extend along cephalic wall from apical spine to sides of vertical spine; four collar pores. Postcephalic segments increase gradually in length and width to largest, generally the fifth segment. Surface smooth except for slightly roughened thorax and occasional specimens with blunt nodes at angles of weak pore-frames on median segments. Strictures between segments well-defined externally. Pores small, circular to subcircular, increasing in size gradually distally. There is a marked tendency toward longitudinal alignment of pores on median post-thoracic segments, although this feature varies considerably from specimen to specimen. On some specimens pores of abdominal segments subdivided.

Length of longest broken specimen of six segments plus a large fragment of the seventh 200 μ, of five segments 105–35 μ, of fifth segment 30–45 μ, of cephalis 12–15 μ, of thorax 15–20 μ; width of thorax 30–4 μ, of fifth segment 65–90 μ. Dimensions based on fifteen specimens.

Discussion. This species is distinguished by its small size, and tendency toward longitudinal alignment of abdominal pores. It differs from *Eucyrtidium montiparum* Ehrenberg (1873) in having a smaller thorax and abdomen.

Localities. CAS loc. 1144, Marca shale (MBP), and Moreno Gulch.

Holotype. USNM 158010, N38/0, from upper Maestrichtian Moreno formation, Marca shale (MBP), Fresno County, California.

Stichomitra asymbatos sp. nov.

Plate 8, figs. 10*a–c*

Description. Shell of six to ten segments, conical except for last one or two which narrow and are subcylindrical. Cephalis with sturdy horn smooth proximally, generally with a few thorns distally or a slightly roughened tip; rarely the horn poorly developed as a thickened apex with a few, short, irregular protrusions. Cephalis either smooth hyaline or with shallow depressions and a few thorns apically; a tiny vertical horn frequently present where the internal vertical spine penetrates the cephalic wall. A single row of inward- and downward-directed pores almost always present at collar stricture.

Internally, four collar pores, the vertical spine (which divides the cervical pore on many of the cyrtoids) rising so sharply that its distal end is markedly above the collar stricture. Internal apical spine prominent along dorsal cephalic wall. Thorax generally has pores or depressions quincuncially arranged, with angular frames and tiny nodes or spines at the angles; rare specimens have no pores or depressions, surface smooth except for a few irregular thorns. A single row of inward- and downward-directed pores present at lumbar stricture. Conical abdominal segments thick-walled, distal subcylindrical ones thin-walled. The thick-walled abdominal segments generally increase gradually in width and length, with the last, or next to last the widest and longest. Thin-walled abdominal segments very variably developed narrowing distally, the terminal one generally considerably longer than any of the other abdominal segments. Length of segments and depth of strictures vary considerably from specimen to specimen. Pores subcircular, rather regular in size on individuals but vary considerably among specimens, either irregularly arranged or tending toward a quincuncial arrangement with transverse rows. Surface rough with walls thickened by large blunt nodes or spines. These appear first on the third segment and are prominent, depending on length of specimen, to fourth or eighth segment. The one or two thin-walled segments generally have no nodes, and only occasionally downwardly directed spines on the proximal segment; the last segment, always incomplete, occasionally shows a portion of its distal edge with irregular downwardly directed teeth.

Length of well-developed apical horn 25–60 μ, of cephalis 22–30 μ, of thorax 22–30 μ, of five segments 140–210 μ, of seven segments 202–310 μ, of fifth segment 30–60 μ, of seventh segment 38–110 μ; width of apical horn near base 10–18 μ, of thorax 43–50 μ, of fifth segment 75–130 μ, of seventh segment 85–175 μ; ratio of width to length of thick-walled fifth segments 1·9–2·5, of thick-walled seventh segments 2·1–3·5. Dimensions based on thirty specimens.

Discussion. The larger size of the cephalis, high rising vertical spine, and sturdy apical horn make it not very likely that this species is closely related to the other two species here included under *Stichomitra.* However it seems even less likely to be related to the type species of *Eucyrtidium.*

There is in the Upper Cretaceous a wide spread distinctive group of stichocyrtids characterized by their large size; spherical poreless cephalis with sturdy horn and internally sharply rising vertical spine; conical proximal segments with ornamentation varying from tiny spines to large blunt nodes, with pores generally irregularly arranged; and cylindrical, narrowing, thinner, distal segments. The very variable species described above represents the upper Maestrichtian member of this group. It may be that when more material is studied with distal segments well-preserved *S. asymbatos* will be divided. At present examination of many specimens, most without thinner distal segments, has shown intergradations permitting no other interpretation than that this is a species varying greatly in size, shape of segments, and surface ornamentation.

Previously described members of this group are *Cyrtophormis (Acanthocyrtis) grandis* Campbell and Clark, *Eucyrtidium Eucyrtis carnegiense* Campbell and Clark, and *E. E. carnegiense* var. *positasense* Campbell and Clark. *Dictyomitra scalaris* Lipman, *D. gigantea* Lipman, and *D.? nodosa* Koslova are also probably closely related. Undescribed species have been observed in Cuba B191, and Trinidad well Marac 1.

Localities. CAS loc. 1144, Marca shale (MBP), Moreno Gulch, Cima Hill hole III (at approximately 306·5, 310, 313·5, 320, 323·5, 327·5, 330, 340, 360, 363·5, and 383 ft.), CAS loc. 39545, UCMP loc. A2615, Stanford loc. PL2875, Evitt slide AE27, and Lamont V–18–129.

Type specimens. Holotype USNM 158013, O38/0 CAS loc. 1144; paratypes USNM 158012, L38/3 CAS loc. 1144, and USNM 158013, O37/3 Moreno Gulch (3518), all from upper Maestrichtian Moreno formation, Fresno County, California.

? *Stichomitra campi* (Campbell and Clark) emend.

Plate 8, figs. 3*a–c*

1944*b* *Cyrtocapsa (Cyrtocapsoma) campi* Campbell and Clark, p. 43, pl. 8, figs. 14, 15, 17, 20, (fig. 16?).

Description. Shell of four to eight segments, conical except for the last one or two segments which may be cylindrical the last one constricted distally, inverted hemispherical. Aperture a large pore or very short tube, its margin smooth or roughened by protruding intervening bars of marginal pores. Cephalis with short, smooth apical horn and tiny thorn-like vertical horn. Sometimes secondary lateral spines extend upward as small spines at the collar stricture. Internally, cephalis divided by paired ridges along cephalic wall between apical and vertical spines. These ridges vary in their position in the cephalis; sometimes near the apex and appear as little more than two lines on the inner surface; generally they are situated somewhat further down constricting the cephalis so that it is divided into three lobes, an apical lobe above the ridges and two lateral lobes below. Collar stricture not well-defined externally, and collar structure of four large collar pores depressed in thorax. Both cephalis and thorax with small, circular, irregularly spaced pores; small dorsal and primary lateral wings sometimes present on thorax. Strictures between post-thoracic segments not pronounced, and segments increase in length only slightly except for last which is very variable in length; the last stricture frequently poorly defined. Pores of abdominal segments very variable. On individual specimens they may be small, circular, uniform in size and arranged quincuncially in vague transverse rows, or they may be variable in size and irregularly distributed. Some may be situated on strictures or, when in approximately transverse rows on adjoining segments, apposed and sometimes coalesced. Surface smooth or with tiny spines.

Length of complete specimens of four to eight segments (exclusive of horn and basal neck) 100–237 μ; length of horn 3–30 μ, of cephalis 15–22 μ, of cephalis and thorax together 30–5 μ, of last segment 25–50 μ; width of widest segment 55–115 μ, of aperture 8–18 μ. Dimensions based on thirty specimens measured not only from the Fresno County, California localities, but also from CAS loc. 39545, and Lamont V–18–129.

Discussion. The specimens described here agree well with specimens of *Cyrtocapsa (Cyrtocapsoma) campi* Campbell and Clark (1944*b*) from the type locality even though Campbell and Clark's description does not mention, and the illustrations do not show the character of the cephalis.

Localities. CAS loc. 1144, Marca shale (MBP), Moreno Gulch, Cima Hill hole III (at approximately 310 and 360 ft.), UCMP loc. A2615, CAS loc. 39545, Stanford loc. PL2875, Evitt slides, and Lamont V–18–129.

Illustrated specimens. USNM 158014, M39/3, USNM 158015, M38/0, and USNM 158016, M38/1, all from upper Maestrichtian Moreno formation, Moreno Gulch (3517), Fresno County, California.

? *Stichomitra livermorensis* (Campbell and Clark) emend.

Plate 8, figs. 2*a, b*

1944*b* *Artocapsa livermorensis* Campbell and Clark, p. 45, pl. 8, figs. 10, 19, 21, 27.

Description. Shell of five to ten segments, conical, except for last segment which closes and may be inverted hemispherical or conical, with a ridged spine basally. Cephalis with short, variably developed, sharp, smooth, apical and vertical horns. Generally secondary lateral spines extend upward as small spines at the collar stricture. Internally, cephalis divided by paired ridges along cephalic wall between apical and vertical spines. These vary in their position in the cephalis but are most frequently well down constricting it to form a distinct apical lobe and two lateral lobes below. Collar stricture not well-defined externally and collar structure with four large collar pores depressed in thorax. Cephalis, except for apical lobe which frequently lacks pores, with small, circular, irregular spaced pores. Thoracic pores similar to those of cephalis. Small dorsal and primary lateral wings sometimes present on thorax. Post-thoracic strictures marked by little or no change in contour; the generally thick wall smooth, except on some specimens which have a few slender scattered spines directed outward medianly and downward distally. Segments increase in length only slightly distally except the last segment which is very variable in size and may be elongated into a short tube or inverted cone just above the basal spine. Abdominal pores on individual specimens may be small, circular, rather uniform throughout; or larger, circular to subcircular, somewhat variable in size. Distribution may be irregular, or quincuncial in vague or distinct transverse rows. Some may be situated on strictures or, when in transverse rows, on adjoining segments apposed or coalesced.

Length of complete specimens of five to nine segments (exclusive of apical horn and basal spine) 118–300 μ, of horn 6–25 μ, of cephalis 15–25 μ, of cephalis and thorax together 30–7 μ, of last segment 23–105 μ, of basal spine 5–50 μ; width of widest segment 55–110 μ. Dimensions based on twenty-five specimens measured from the three Fresno County localities, and CAS loc. 39545, and Lamont V–18–129.

Discussion. This species agrees well with specimens of *Artocapsa livermorensis* Campbell and Clark (1944*b*) from the type material except that the specimens described above commonly have a thick wall, although all variations between thick and thin are present. The description for *A. livermorensis*, however, makes no mention of the unusual cephalic structure.

Localities. CAS loc. 1144, Marca shale (MBP), Moreno Gulch, Cima Hill hole III (at approximately 306·5, 310, 313·5, 330, 340, and 360 ft.), CAS loc. 39545, UCMP loc. A2615, Stanford loc. PL2875, Evitt slides, and Lamont V–18–129.

Illustrated specimens. USNM 158017, N39/0 and USNM 158018, N38/1, both from upper Maestrichtian Moreno formation, Moreno Gulch (3517), Fresno County, California.

? *Stichomitra cechena* sp. nov.

Plate 8, fig. 1

Description. Shell of seven to nine, possibly more, segments; conical proximally, cylindrical distally; last segment narrowed. Cephalis, divided into three well-developed lobes as in ?*S. campi* and ?*S. livermorensis*, bears short smooth apical horn, tiny vertical horn and secondary lateral spines extending upward at the collar stricture. Two lateral lobes and rarely apical lobe have small circular, irregularly spaced pores as does the thorax. Collar stricture not well-defined externally, collar structures partly depressed in thorax. Three or four conical post-thoracic segments inflated, increasing gradually in length and width while cylindrical segments, except the last, are more uniform, less inflated. Last segment very variable in length with margin of irregular teeth. Pores of post-thoracic conical segments small, circular, arranged irregularly or quincuncially with only a slight tendency towards transverse alignment. On the distal cylindrical segments they may become larger and more irregular in size and arrangement. On at least some of the proximal segments, pores recessed in indistinct frames with blunt nodes or tiny thorns at the corners. Long, slender, outward- and downward-directed by-spines sparsely and irregularly distributed over median and distal segments.

Length of longest specimen of nine segments 306 μ, of seven segments 158–232 μ, of cephalis 20–5 μ, of cephalis and thorax together 40 μ, of widest segment 25–45 μ (most 40 μ), of apical horn 15–25 μ of long by-spines 12–25 μ; width of thorax 35–45 μ, of widest segment 70–100 μ. Dimensions based on thirteen specimens.

Discussion. This species differs from the other two described species of ? *Stichomitra* with a similar cephalis by its open aperture and long, slender by-spines on distal portion of shell.

Localities. CAS loc. 1144, Marca shale (MBP), and Moreno Gulch.

Holotype. USNM 158019, Q38/3 from upper Maestrichtian Moreno formation, Marca shale (MBP), Fresno County, California.

? *Stichomitra alamedaensis* (Campbell and Clark) emend.

Plate 8, fig. 4

1944*b* *Phormocampe* (*Cyrtocorys*) *alamedaensis* Campbell and Clark, p. 37, pl. 7, fig. 41.
?1944*b* *Phormocampe* (*Cyrtocorys*) *alamedaensis* var. *tenuis* Campbell and Clark, p. 37, pl. 7, figs. 35, 36.

Description. Shell of up to nine segments, possibly more, conical, the last preserved segment frequently constricted; its distal margin ragged, broken. Cephalis with upper two-thirds almost spherical, poreless, hyaline, with tiny conical nodes connected by bars to form a triangular pattern. At its base a ring similar to that in *Amphipyndax*; collar structure of ? four large pores below this ring within thorax, attached to ring by apical and vertical spines. Rarely, tiny apical and vertical horns present on the sphere. No collar stricture, but a distinct stricture where the spherical part of the cephalis and the thorax join. Thorax has closely and irregularly spaced, small, circular, pores. Post-thoracic segments increase gradually in length and width, each stricture well indented, surface smooth. Pores circular, closely spaced, in weak angular frames, arranged quincuncially in transverse rows, three to five rows per segment, rarely irregular proximally. Pores rather uniform in size on a segment, gradually increasing in size distally; on adjoining segments apposed and frequently coalesced.

Length of longest broken specimen of eight segments plus fragment of ninth 185 μ, of five segments 90–100 μ, of spherical part of cephalis 16–18 μ, of complete cephalis 22–5 μ, of thorax 17–20 μ; greatest width of sixth segment 70–80 μ. Dimensions based on fifteen specimens.

Discussion. This species is uniform in its development and easily recognizable by its unusual cephalis and collar structures in thoracic cavity. A specimen from the lower Maestrichtian is almost closed basally, with only a large terminal pore. This is the only one of the described species which is common throughout the probable Paleocene of Cima Hill core hole III.

There are many species in the literature superficially similar to ? *S. alamedaensis* but because details of the cephalis are not given it is not possible confidently to synonymize or separate them. It is conspecific with *Phormocampe* (*Cyrtocorys*) *alamedaensis*, but a bubble around the collar structures of *P. alamedaensis* var. *tenuis* (Campbell and Clark 1944*b*, pl. 7, fig. 36) makes a certain identification with that variety impossible. Figured specimen 35 was not found. A superficially similar species, as yet undescribed, differs in having the more normal small four collar pores at the stricture and a smaller thorax.

Although the cephalic structure exhibits some similarity to *Amphipyndax*, in that the apical-vertical ridges are developed to the point of forming a ring the main collar structures seem more probably related to that of the other species assigned herein to ? *Stichomitra*. Resolution of this question must await further detailed, comparative morphological studies.

Localities. CAS loc. 1144, Marca shale (MBP), Moreno Gulch, Cima Hill hole III (at approximately 178·5, 180·5, 193·5, 214, 252, 264·5, 271·5, 275·5, 280·5, 286, 287, 291·5, 295·5, and 306·5 ft.), CAS loc. 39545, UCMP loc. A2615, and Stanford loc. PL2875.

Illustrated specimen. USNM 158020, N38/3 Moreno Gulch (3517), from upper Maestrichtian Moreno formation, Fresno County, California.

Genus *Amphipyndax* Foreman 1966

Amphipyndax stocki (Campbell and Clark) emend.

Plate 8, figs. 12*a–c*

1944*b* *Stichocapsa* (?) *stocki* Campbell and Clark, p. 44, pl. 8, figs. 31, 32, 33.
1944*b* *Stichocapsa megalocephalia* Campbell and Clark, p. 44, pl. 8, figs. 26, 34.
?1965 *Dictyomitra uralica* Gorbovetz, in Koslova and Gorbovetz, p. 116, pl. 6, figs. 6, 7.

Description. Shell of five to twelve segments, possibly more. Cephalis knob-like, constricting to form a neck, poreless with numerous nodes or short spiny projections apically, lower half smooth. Internally, cephalis divided about two-thirds down by the ring typical of this genus; four collar pores basally. Thorax and proximal abdominal segments conical, median segments conical or cylindrical and last one or two, rarely three, inversely conical, the last constricting rapidly, sometimes forming a short tube; aperture small, circular with ragged edge. Abdominal segments increase very little in length except for last segment which may be considerably longer than the others. Pores of thorax small, circular, irregularly arranged; surface smooth, rarely with nodes. Post-thoracic segments, except for distalmost, with pores arranged quincuncially in transverse rows, three to four, rarely five rows per segment. Pores at either side of strictures apposed and frequently coalesced. Size of pores and surface ornamentation varies greatly. Individuals with small pores tend to have pores of uniform size evenly distributed, while specimens with larger pores show greater variation in their size and arrangement. Surface may be smooth with slight or no external strictures, so that the specimen appears almost bottle-shaped, or pores of the median segments may be surrounded by hexagonally distributed nodes and rarely by hexagonal frames. Specimens with nodes generally also have pronounced strictures externally.

Length of longest specimen of ten segments plus a fragment 326 μ, of shortest complete specimen of five segments 138 μ, of five segments 115–45 μ, of eight segments 212–72 μ, of cephalis 35–40 μ, of thorax 15–20 μ, of fifth segment 25–38 μ; width (exclusive of nodes) of thorax 35–50 μ, of fifth segment 77–100 μ, of widest segment 85–120 μ. Dimensions based on twenty specimens. Length of two complete specimens of eight and nine segments from Lamont V–18–129, 350 μ each.

Discussion. The variable features all intergrade, and although individual specimens vary considerably in over all appearance no subdivision of the species seems possible. The species is distinguished from *A. plousios* as described under that species.

Gorbovetz in (Koslova and Gorbovetz 1966) recorded a similar species (only doubtfully synonymized because no topotypic material is available for study) from the Campanian of western Siberia, upper part of the Berezovskay horizon of the Tumensky–Petropavlovsky, Berezov–Tuhuelsky regions, and the western slope of the Urals.

Localities. CAS loc. 1144, Marca shale (MBP), Moreno Gulch, Cima Hill hole III (at approximately 275·5, 313·5, 320, and 383 ft.), CAS loc. 39545, UCMP loc. A2615, Stanford loc. PL2875, Lamont V–18–129, Cuba B191, and Trinidad well Marac 1.

Illustrated specimens. USNM 158201, O37/1 Marca shale (MBP), USNM 158022, N38/0 Moreno Gulch (3518), and USNM 158023, K40/4 CAS loc. 1144, all from upper Maestrichtian Moreno formation, Fresno County, California.

Amphipyndax plousios sp. nov.

Plate 8, fig. 11

Description. Shell of six to nine segments, possibly more. Cephalis knob-like, constricting to form a neck, poreless with numerous nodes or short spiny projections apically, lower half smooth. Internally, cephalis divided about two-thirds down by ring; four collar pores. Thorax and most abdominal segments conical, last but one may be annular, and last observed segment, never seen complete, constricting with a ragged lower margin. Abdominal segments increase regularly in length except for the last two or three which frequently are markedly longer. Pores small, circular to subcircular, irregularly arranged, approximately uniform in size on each segment and only very slightly, if at all, larger distally. Most specimens with median segments roughened by nodes.

Length of longest specimen of eight segments plus a fragment 346 μ, of five segments 148–65 μ, of cephalis 35 μ, of thorax 15–20 μ, of fifth segment 37–55 μ; width (exclusive of nodes) of thorax 35–50 μ, of fifth segment 85–105 μ, of widest segment 85–135 μ. Dimensions based on twelve specimens.

Discussion. This species differs from *A. stocki* in having fewer and longer segments, with small pores irregularly arranged.

Localities. CAS loc. 1144, Marca shale (MBP), Moreno Gulch, Cima Hill hole III (at approximately 313·5, 320, and 383 ft.), CAS loc. 39545, and Evitt slide AE26.

Holotype. USNM 158024, M38/0 from upper Maestrichtian Moreno formation, Marca shale (MBP), Fresno County, California.

REFERENCES

ALIEV, KH. SH. 1965. Radiolarians of the Lower Cretaceous deposits of northeastern Azerbaidzhan and their stratigraphic significance. *Izdatelstvo Akad. Nauk Azerb.* SSR, Baku, 124 pp., pl. 1–15. [In Russian.]

BOLIN, E. J. 1956. Upper Cretaceous Foraminifera, Ostracoda, and Radiolaria from Minnesota. *J. Paleont.* **30**, 278–98, pl. 37–9.

BOLLI, H. M. 1957. Planktonic Foraminifera from the Oligocene-Miocene Cipero and Lengua formations of Trinidad, B. W. I. *Bull. U.S. Nat. Hist. Mus.* **215**, 97–123, pl. 22–9.

BORISENKO, N. N. 1958. Paleocene Radiolaria of western Kuban *in* Krylov, A. P., ed., Questions on geological boring and the exploitation of wells. *Trudȳ vses. neftegas. nauchno-issled. Inst. VNll*, Krasnodarskii Filial **17**, 81–100, pl. 1–6. [In Russian.]

BURMA, B. H. 1959. On the status of *Theocampe* Haeckel and certain similar genera. *Micropaleontology*, **5**, 325–30.

BURY, P. S. 1862. *Figures of remarkable forms of polycystins, or allied organisms, in the Barbadoes chalk deposits.* 4 pp., 26 pl. [London.]

BÜTSCHLI, O. 1882. Beiträge zur Kenntnis der Radiolarienskelette, insbesondere der der Cyrtida. *Z. wiss. Zool.* **36**, 485–540, pl. 31–3.

CAMPBELL, A. S. and CLARK, B. L. 1944a. Miocene radiolarian faunas from Southern California. *Sp. Pap. geol. Soc. Amer.* **51**, 1–76, pl. 1–7.

—— and —— 1944b. Radiolaria from Upper Cretaceous of Middle California. Ibid. **57**, 1–61, pl. 1–8.

CARNEVALE, P. 1908. Radiolarie e Silicoflagellati di Bergonzano (Reggio Emilia). *Mem. R. Ist. Veneto Sci. Lett. Arti* **28**(3), 1–46, pl. 1–4.

CAYEUX, L. 1897. Contribution à l'étude micrographique des terrains sédimentaires. *Mém. Soc. Géol. Nord, Lille* **4**(2), 1–591, pl. 1–10.

DEFLANDRE, G. 1953. *Radiolaires fossiles* in Traité de Zoologie (ed. P.-P. Grassé, Masson, Paris) **1**(2), 389–436.

DREYER, F. 1889. Morphologische Radiolarienstudien. Heft 1, Die Pylombildungen in vergleichend-anatomischer und entwicklungsgeschichtlicher Beziehung bei Radiolarien und bei Protisten überhaupt, nebst System und Beschreibung neuer und der bis jetzt bekannten pylomatischen Spumellarien. *Jena. Z. Naturw.* **23**, N.F. 16, 1–138, pl. 1–6.

DRUGG, W. S. 1967. Palynology of the Upper Moreno Formation (late Cretaceous-Paleocene) Escarpado Canyon, California. *Palaeontographica*, Abt. B, **118**, 1–71, pl. 1–9.

EHRENBERG, C. G. 1838. Über die Bildung der Kreidefelsen und des Kreidemergels durch unsichtbare Organismen. *Abh. preuss. Akad. Wiss. Berlin*, Jahrg. 1838, 59–147, pl. 1–4.

—— 1844. Über 2 neue Lager von Gebirgsmassen aus Infusorien als Meeres-Absatz in Nord-America und eine Vergleichung derselben mit den organischen Kreide-Gebilden in Europa und Afrika. *Mber. preuss. Akad. Wiss. Berlin*, Jahrg. 1844, 57–97.

—— 1847*a*. Über eine halibiolithische, von Herrn R. Schomburgk entdeckte, vorherrschend aus mikroskopischen Polycystinen gebildete Gebirgsmasse von Barbados. Ibid., Jahrg. 1846, 382–85.

—— 1847*b*. Über die mikroskopischen kieselschaligen Polycystinen als mächtige Gebirgsmasse von Barbados und über das Verhältniss der aus mehr als 300 neuen Arten bestehender ganz eigenthümlichen Formengruppe jener Felsmasse zu der jetzt lebender Thieren und zur Kreidebildung. Eine neue Anregung zur Erforschung des Erdlebens. Ibid., Jahrg. 1847, 40–60, pl. 1.

—— 1854. Die systematische Charakteristik der neuen mikroskopischen Organismen des tiefen atlantischen Oceans für den Monatsbericht zum Druck zu übergeben, deren Verzeichniss im Monat Februar bereits mitgetheilt worden ist. Ibid., Jahrg. 1854, 236–50.

—— 1873. Grössere Felsproben des Polycystinen-Mergels von Barbados mit weiteren Erläuterungen. Ibid., Jahrg. 1873, 213–63.

—— 1875. Fortsetzung der mikrogeologischen Studien als Gesammt-Uebersicht der mikroskopischen Paläontologie gleichartig analysirter Gebirgsarten der Erde, mit specieller Rücksicht auf den Polycystinen-Mergel von Barbados. *Abh. preuss. Akad. Wiss. Berlin*, Jahrg. 1875, 1–226, pl. 1–30.

EICHER, P. L. 1960. Stratigraphy and micropaleontology of the Thermopolis shale. *Bull. Peabody Mus. Nat. Hist., Yale Univ.* **15**, 1–126, pl. 1–6.

FOREMAN, H. P. 1966. Two Cretaceous radiolarian genera. *Micropaleontology*, **12**, 355–59.

FRIZZELL, D. L. and MIDDOUR, E. S. 1951. Paleocene Radiolaria from southeastern Missouri. *Bull. Univ. Mo. Sch. Min. Metall., Tech. Ser.* **77**, 1–41, pl. 1–3.

GALAVIS, S. F. 1951. Los organismos silíceos y sus posibles usos en métodos correlativos. *Boln Geol. Minist. Minas Venez.* 1 (3), 313–24, pl. 1.

GIGNOUX, M. 1955. *Stratigraphic geology* (English translation from 4th French ed. 1950), San Francisco.

GOUDKOFF, P. P. 1945. Stratigraphic relations of Upper Cretaceous in Great Valley, California. *Bull. Am. Ass. Petrol. Geol.* **29**, 956–1007.

HAECKEL, E. 1860. Abbildungen und Diagnosen neuer Gattungen und Arten von lebenden Radiolarien des Mittelmeeres. *Mber. preuss. Akad. Wiss. Berlin*, Jahrg. 1860, 835–45.

—— 1862. *Die Radiolarien. (Rhizopoda Radiaria.). Eine Monographie*, Berlin.

—— 1881. Entwurf eines Radiolarien-Systems auf Grund von Studien der Challenger-Radiolarien. *Jena. Z. Naturw.* **15**, N.F. 8(3), 418–72.

—— 1887. Report on the Radiolaria collected by H. M. S. Challenger during the years 1873–76. *Rept. Voyage Challenger*, Zool. **18**, clxxxviii+1803, pl. 1–140.

HANNA, G. D. 1928. Silicoflagellata from the Cretaceous of California. *J. Paleont.* **1**, 259–63, pl. 41.

HINDE, G. J. 1900. Description of fossil Radiolaria from the rocks of Central Borneo. *in* Molengraaff, G. A. F., *Borneo-Expedite: Geologische Verkenningstochten in Centraal-Borneo (1893–1894)*, Brill, Leiden. Appendix 1, 1–51, pl. 1–4.

—— 1908. Radiolaria from Triassic and other rocks of the Dutch East Indian Archipelago. *Jaarb. Mijnw. Ned.-Oost-Indië*, **37**, 1–42, pl. 5–10.

HOLMES, W. M. 1900. On Radiolaria from the Upper Chalk at Coulsdon (Surrey). *Q. Jl geol. Soc. Lond.* **56**, 694–704, pl. 37–8.

HÜLSEMANN, K. 1963. Radiolaria in plankton from the Arctic drifting station T-3, including the description of three new species. *Arct. Inst. N. Am., Tech. Pap.* **13**, 1–52.

IACOB, D. and NICORICI, E. 1957. Citeva forme de Radiolari din jaspurile de la Tămăşeşti (Reg. Hunedoara). *Studii Cerc. Geol.-Geogr. Cluj*, anul 8 (-2), 7–19.

KOSLOVA, G. E. and GORBOVETZ, A. N. 1966. Radiolaria of the Upper Cretaceous and Upper Eocene deposits of the West Siberian Lowland. *Trudy vses. neft. nauchno-issled. geol.-razv. Inst.* **248**, 1–159, pl. 1–17. [In Russian.]

LIPMAN, R. H. 1952. Materials for the monographic study of Radiolaria of the Upper Cretaceous deposits of the Russian Platform. *Paleontologiya i Stratigrafiya (Trudy vses. nauchno-issled. geol. Inst.)* 24–51, pl. 1–3. [In Russian.]

—— 1960. Radiolaria *in* Stratigraphy and fauna of the Cretaceous deposits of the Western Siberian Plain. *Trudy vses. nauchno-issled. geol. Inst.*, new ser. **29**, 124–34, pl. 26–32. [In Russian.]

—— 1961. Sum-total of studies of the Late-Cretaceous and Paleogene Radiolaria in the Western Siberian Lowlands, Russian Platform, also Central Asia. *Vsesoyuznoe Paleontologicheskoe Obshchestvo, Sorok Let Sovetskoi Paleontologii (1917–1957)*, Trudy 4th session, 41–6. [In Russian.]

LONG, J. A., FUGE, D. P., and SMITH, J. 1946. Diatoms of the Moreno shale. *J. Paleont.* **20**, 89–118, pl. 13–19.

NAKASEKO, K. 1963. Neogene Cyrtoidea (Radiolaria) from the Isozaki Formation in Ibaraki Prefecture, Japan. *Osaka Univ. Sci. Repts.* **12**(2), 165–98, pl. 1–4.

NAUSS, A. W. 1947. Cretaceous microfossils of the Vermilion area, Alberta. *J. Paleont.* **21**, 329–43, pl. 48, 49.

NEVIANI, A. 1900. Supplemento alla fauna a Radiolari delle rocce Mesozoiche del Bolognese. *Boll. Soc. geol. ital.* **19**, 645–71, pl. 9, 10.

NIGRINI, C. A. 1967. Radiolaria in pelagic sediments from the Indian and Atlantic Oceans. *Bull. Scripps Inst. Oceanogr.*, La Jolla, **11**, 1–106, pls. 1–9.

PERNER, J. 1891. O radiolariích z českého útvaru křídového. *Věst. Král. české spol. nauk.* Třída math.-přír. (*Sber. K. böhm. Ges. Wiss.*) 255–69, pl. 10.

PESSAGNO, E. A. 1963. Upper Cretaceous Radiolaria from Puerto Rico. *Micropaleontology*, **9**, 197–214, pl. 1–7.

PETRUSHEVSKAYA, M. G. 1964. On homologies in the elements of the inner skeleton of some Nassellaria. *Zool. Zh.* **43**(8), 1121–8. [In Russian.]

—— 1965. Peculiarities of the construction of the skeleton of Botryoids (Order Nassellaria). *Trudȳ zool. Inst. Leningr.* **35**, 79–118. [In Russian.]

POPOFSKY, A. 1913. Die Nassellarien des Warmwassergebietes. *Dt. Südpol.-Exped.* 10, *Zool.* **6**, 217–416, pl. 28–38.

PRINCIPI, P. 1909. Contributo allo studio dei Radiolari Miocenici Italiani. *Boll. Soc. geol. ital.* **28**, 1–22, pl. 1.

RICHTER, M. 1925. Beiträge zur Kenntnis der Kreide in Feuerland. *Neues Jb. Miner. Geol. Paläont.* BeilBd. **52**, Abt. B, 524–568, pl. 6–9.

RIEDEL, W. R. 1957. Radiolaria: a preliminary stratigraphy. *Rep. Swed. deep Sea Exped.* 6(3), 59–96, pl. 1–4.

—— 1959. Oligocene and Lower Miocene Radiolaria in tropical Pacific sediments. *Micropaleontology*, **5**, 285–302, pl. 1–2.

RÜST, D. 1885. Beiträge zur Kenntniss der fossilen Radiolarien aus Gesteinen des Jura, *Palaeontographica*, **31**, ser. 3, v. **7**, 269–321, pl. 26–45.

—— 1892. Radiolaria from the Pierre formation of North-Western Manitoba. *Contr. Can. Micro-Pal.*, *geol. Nat. Hist. Surv. Can.*, pt. 4, 101–10, pl. 14–16.

—— 1898. Neue Beiträge zur Kenntniss der Fossilen Radiolarien aus Gesteinen des Jura und der Kreide. *Palaeontographica*, **45**, 1–67, pl. 1–19.

SQUINABOL, S. 1903. Le Radiolarie dei noduli selciosi nella scaglia degli Euganei. *Riv. ital. Paleont.* **9**, 105–50, pl. 8–10.

—— 1904. Radiolarie Cretacee degli Euganei. *Atti Memorie Accad. Sci. Lett.*, *Padova*, new ser. **20**, 171–244, pl. 1–10.

—— 1914. Contributo alla conoscenza dei Radiolarii fossili del Veneto. *Memorie Ist. Univ. Padova*, **2**, 249–306, pl. 20–4.

STÖHR, E. 1880. Die Radiolarienfauna der Tripoli von Grotte Provinz Girgenti in Sicilien. *Palaeontographica*, **26**, ser. 3, v. 2, 69–124, pl. 17–23.

TAN SIN HOK. 1927. Over de samenstelling en het onstaan van krijt- en mergelgesteenten van de Molukken. *Jaarb. Mijnw. Ned.-Oost-Indië*, jaarg. 1926, Verhand., pt. 3, 5–165, pl. 1–16.

ZITTEL, K. A. 1876. Ueber einige fossile Radiolarien aus der norddeutschen Kreide. *Dt. geol. Ges. Z.* **28**, 75–86, pl. 2.

HELEN P. FOREMAN
Department of Geology,
Oberlin College,
Oberlin, Ohio 44074,
U.S.A.

Typescript received 17 April 1967